Thinking Probabilistically

Probability theory has diverse applications in a plethora of fields, including physics, engineering, computer science, chemistry, biology, and economics. This book will familiarize students with various applications of probability theory, stochastic modeling, and random processes, using examples from all these disciplines and more.

The reader learns via case studies and begins to recognize the sort of problems that are best tackled probabilistically. The emphasis is on conceptual understanding, the development of intuition, and gaining insight, keeping technicalities to a minimum. Nevertheless, a glimpse into the depth of the topics is provided, preparing students for more specialized texts while assuming only an undergraduate-level background in mathematics. The wide range of areas covered – never before discussed together in a unified fashion – includes Markov processes and random walks, Langevin and Fokker–Planck equations, noise, generalized central limit theorem and extreme values statistics, random matrix theory, and percolation theory.

Ariel Amir is a Professor at Harvard University. His research centers on the theory of complex systems.

Thinking Probabilistically

Stochastic Processes, Disordered Systems, and Their Applications

ARIEL AMIR

Harvard University, Massachusetts

CAMBRIDGE
UNIVERSITY PRESS

CAMBRIDGE
UNIVERSITY PRESS

University Printing House, Cambridge CB2 8BS, United Kingdom

One Liberty Plaza, 20th Floor, New York, NY 10006, USA

477 Williamstown Road, Port Melbourne, VIC 3207, Australia

314–321, 3rd Floor, Plot 3, Splendor Forum, Jasola District Centre, New Delhi – 110025, India

79 Anson Road, #06–04/06, Singapore 079906

Cambridge University Press is part of the University of Cambridge.

It furthers the University's mission by disseminating knowledge in the pursuit of education, learning, and research at the highest international levels of excellence.

www.cambridge.org
Information on this title: www.cambridge.org/9781108479523
DOI: 10.1017/9781108855259

First published 2021

Printed in the United Kingdom by TJ Books Ltd. Padstow, Cornwall

A catalogue record for this publication is available from the British Library.

Library of Congress Cataloging-in-Publication Data
Names: Amir, Ariel, 1981– author.
Title: Thinking probabilistically : stochastic processes, disordered systems, and their applications / Ariel Amir.
Description: Cambridge, United Kingdom ; New York, NY : Cambridge University Press, 2021. | Includes bibliographical references and index.
Identifiers: LCCN 2020019651 (print) | LCCN 2020019652 (ebook) | ISBN 9781108479523 (hardback) | ISBN 9781108789981 (paperback) | ISBN 9781108855259 (epub)
Subjects: LCSH: Probabilities–Textbooks. | Stochastic processes–Textbooks. | Order-disorder models–Textbooks.
Classification: LCC QA273 .A548 2021 (print) | LCC QA273 (ebook) | DDC 519.2–dc23
LC record available at https://lccn.loc.gov/2020019651
LC ebook record available at https://lccn.loc.gov/2020019652

ISBN 978-1-108-47952-3 Hardback
ISBN 978-1-108-78998-1 Paperback

Contents

Acknowledgments

I am indebted to the students of Harvard course APMTH 203 for their patience and perseverance as the course materials were developed, and I am hugely grateful to all of my teaching fellows along the years for their hard work: Sarah Kostinksi, Po-Yi Ho, Felix Wong, Siheng Chen, Pétur Rafn Bryde, and Jiseon Min. Much of the contents of the Appendices draws on their helpful notes.

I thank Eli Barkai, Stas Burov, Ori Hirschberg, Yipei Guo, Jie Lin, David Nelson, Efi Shahmoon, Pierpaolo Vivo, and Ahmad Zareei for numerous useful discussions and comments on the notes. Christopher Bergevin had excellent suggestions for Chapter 2, and Julien Tailleur for Chapter 3. Ori Hirschberg and Eli Barkai had many important comments and useful suggestions regarding Chapter 6. I thank Grace Zhang, Satya Majumdar, and Fernando L. Metz for a careful reading of Chapter 8. I am grateful to Martin Z. Bazant and Bertrand I. Halperin for important comments on Chapter 9. I thank Terry Tao for allowing me to adapt the discussion of Black–Scholes in Chapter 3 from his insightful blog. I also thank Dr. Yasmine Meroz and Dr. Ben Golub for giving guest lectures in early versions of the course. Finally, I am grateful to Simon's coffee shop for providing reliably excellent coffee, which was instrumental to the writing of this book.

I dedicate this book to Lindy, Maayan, Tal, and Ella, who keep my feet on the ground and a smile on my face.

1 Introduction

I know too well that these arguments from probabilities are impostors, and unless great caution is observed in the use of them they are apt to be deceptive – in geometry, and in other things too

(from Plato's Phaedo)

The purposes of this book are to familiarize you with a broad range of examples where randomness plays a key role, develop an intuition for it, and get to the level where you may read a recent research paper on the subject and be able to understand the terminology, the context, and the tools used. This is in a sense the "organizing principle" behind the various chapters: In all of them we are driven by applications where probability plays a fundamental role, and leads to exciting and often intriguing phenomena. There are many relations between the chapters, both in terms of the mathematical tools and in some cases in terms of the physical processes involved, but one chapter does not follow from the previous one by necessity or hinge on it – rather, the idea is to present a rich repertoire of problems involving randomness, giving the reader a good basis in a broad range of fields ... and to have fun along the way.

Randomness leads to new phenomena. In a classic paper, Anderson (1972) coined the phrase "more is different". It is also true that "stochastic is different" ... The book will give you some tools to understand phenomena associated with disordered systems and stochastic processes. These will include percolation (relevant for polymers, gels, social networks, epidemic spreading); random matrix theory (relevant for understanding the statistics of nuclear and atomic levels, model certain properties of ecological systems and more); random walks and Langevin equations (pertinent to understanding numerous applications in physics, chemistry, cell biology as well as finance). The emphasis will be on understanding the phenomena and quantifying them. Note that while all of the applications considered here build on randomness in a fundamental way, the collection of topics covered is far from representative of the vast realm of applications that hinge on probability theory. (For instance, two important fields not touched on here are statistical inference and chemical kinetics).

What mathematical background is assumed? The book assumes a solid background in undergraduate-level mathematics: primarily calculus, linear algebra, and basic probability theory. For instance, when a PDE (partial differential equation) is derived, we will not dwell too much on its solution if standard techniques are utilized, as will be the case when using standard results of linear algebra (e.g., that a Hermitian matrix admits unitary diagonalization). Similarly, if Lagrange multipliers are needed to perform a minimization, familiarity with them will be *assumed*. We will occasionally evaluate integrals using contour integration, though a motivated reader will be able to follow the vast majority of the book without a background in complex analysis. A summary/refresher of some of the techniques utilized is provided in Appendices A–E. A concise mathematical physics textbook that may come in handy for a reader who needs a further reminder of the techniques is by Mathews and Walker (1970). For readers who need to fill in a gap in their mathematical background (e.g., complex analysis), two excellent textbooks that cover the vast majority of the mathematical background assumed here (and much more) are by Hassani (2013) and Arfken, Weber, and Harris (2012). Further references for particular subject matter (probability theory, linear algebra, etc.) are provided in the Appendices.

To Prove or Not to Prove … That Is the Question It is also important to emphasize what this book is **not** about: The derivations we will present will *not* be mathematically rigorous. What a physicist considers a proof is not what a mathematician would! In many cases, the physicist's proof has to be later "redone" by a mathematician, sometimes decades later. However, for various real-world applications the advantages of a rigorous proof might not be justifiable. Quoting Feynman, "If there is something very slightly wrong in our definition of the theories, then the full mathematical rigor may convert these errors into ridiculous conclusions." To paraphrase Feynman – in many cases, there is no need to solve a model exactly, since the connection between the model and the reality is only crude, and we made numerous (far more important) approximations in deriving the model equations (von Neumann put this more bluntly: "There's no sense in being precise when you don't even know what you're talking about"). Similar expectations will hold for the exercises at the end of each chapter, which also consist of numerical simulations in cases where analytic derivations are impossible or outside the scope of the book.

Furthermore, the notation and jargon we will follow will be those used by physicists in research papers – which is sometimes different from those used by mathematicians (for instance, we will refer to the object mathematicians call the "probability density function" as the "probability distribution", and use the physicists' notation for it – $p(x)$). We will often not specify the precise mathematical conditions under which the derivations can be made rigorous, and we will shamelessly use algebraic manipulations without justifying them – such as using Fubini's theorem without ensuring the function is absolutely integrable. All functions will be assumed to be differentiable as many times as necessary. We will also be using Dirac's δ-function in the derivations, and creatures such as the Fourier transform of $e^{i\omega t}$. For a physicist, these objects should be interpreted under the appropriate regularization – a δ-function should be thought of as having a finite width, but much smaller than any other relevant scale in the problem. (Physicists are relatively used to this sort of regularization – for instance, in computing Green functions using contour integration the contour often has to be shifted by an amount $\pm i\epsilon$ to make the results convergent). If the final result depends on this finite width – then the treatment using δ-function is inadequate and should be revisited. But as long as the final results are plausible (e.g., in some cases we can compare with numerics) we will not re-derive them in a rigorous fashion. The advantage of this non-rigorous approach is that it seems to be the one more relevant to applications, which are the focus of this book. Experiments and real-life phenomena do not conform to the mathematical idealization we make anyhow, and von Neuman's quote comes to mind again. In other words, the more important thing for explaining physical reality is to have a good model rather than specify the conditions rigorously (a related quote is attributed to Kolmogorov: "Important is not what is rigorous but what is true"). That is not to take anything away from the beautiful work of mathematicians – it is just not the point of this book.

For some students, this non-rigorous approach could prove challenging. When the rigorously inclined student encounters a situation where they feel the formal manipulations are unjustified, it may prove useful for them to construct counter-examples, e.g., functions which do not obey the theorem, and then to consider the physical meaning of these "good" and "bad" functions – which class is relevant in which physical situations, and what we learn from the scenarios where the derivation fails. Our goal here is not to undermine the importance of rigorous mathematics, but to provide a non-rigorous introduction to the plethora of natural sciences phenomena and applications where stochasticity plays a central role.

How to read this book A few words on the different topics covered and their relations. Chapter 1 is introductory, and gives some elementary examples where basic probability theory leads to perhaps counter-intuitive results. One of the examples, Benford's law, touches on some of the topics of Chapter 6 (dealing with heavy-tailed distributions, falling off as a power-law). Chapter 2 presents random walks and diffusion, and provides the foundational basis for many of the other chapters. Chapter 3 directly builds on the simple random walks introduced in Chapter 2, and discusses the important concepts of Langevin and Fokker–Planck equations. The first part of the chapter "builds" the formalism (albeit in a non-technical and non-rigorous fashion), while the second part of the chapter deals with three applications of the ideas (cell size control – an application in biology, the Black–Scholes equation – one in economics, and finally a short application in hydrology). A reader may skip these applications without affecting the readability of the rest of the materials. Similarly, Chapter 4 (dealing with the "escape over a barrier" problem) can be viewed as a sophisticated application of the ideas of Chapter 3, with far-reaching implications. It certainly puts the materials of the previous chapters to good use, but again can be skipped without affecting the flow. Chapter 5 is of particular importance to those dealing with signals and noise, and builds on ideas introduced in earlier chapters (e.g., the Markov chains of Chapter 2) to analyze the power spectrum (i.e., noise characteristics) of several paradigmatic systems (including white noise, telegraph noise, and $1/f$ noise). Chapter 6 derives a plethora of basic results dealing with the central limit theorem, its limitations and generalizations, and the related problem of "extreme value distributions". It is more technical (and lengthier) than previous chapters. Chapter 7, dealing with anomalous diffusion, can be viewed as an advanced application of the materials of Chapter 6, "reaping the fruits" of the labor of the previous chapter. In a sense, it extends the results of the random walks of Chapter 2 to scenarios where some of the assumptions of Einstein's approach do not hold – and have been shown to be relevant to many systems in physics and biology (reminiscent of the quote, "everything not forbidden is compulsory" . . .) Chapter 8 deals with random matrices and some of their applications. It is the most technical chapter in this book, and is mostly independent from the chapter on percolation theory that follows. Moreover, a large fraction of Chapter 8 deals with a non-trivial derivation of the "circular law" associated with non-Hermitian matrices, and a reader can skip directly to Chapter 9 if they prefer. (Note that most of this lengthy derivation "unzips" the short statements made in the original paper, perhaps giving students a glimpse into the compact nature in which modern research papers are written!) The final chapter on percolation theory touches on fundamental concepts such as emergent behavior, the renormalization group, and critical phenomena. Throughout the chapters, numerical simulations in MATLAB are provided when relevant.* Often, results are easy to obtain numerically but challenging to derive analytically, highlighting the importance of the former as a supplement to analytic approaches. Finally, note that occasionally "boxes" are used where we emphasize an idea or concept by placing the passage between two solid lines.

Note that each chapter deals with a topic on which many books and many hundreds of papers have been written. This book merely opens a narrow window into this vast literature. The references throughout the book are also by no means comprehensive, and we apologize for not including numerous relevant references – this text is not intended to be a comprehensive guide to the literature! When possible, we refer to textbooks on the topic that provide a more in-depth discussion as well as a more extensive list of references.

A comment on the problems in this book (and their philosophy) The problems at the end of each chapter are a little different from those encountered in most textbooks. The phrasing is often laconic or even vague. Students might complain that "the problem is not hard – I just cannot figure out what it is!" This actually reflects the typical situation in many real-life problems, be it in academia or industry, where figuring out how to *set up* the problem is often far more challenging than solving the problem itself. The Google PageRank algorithm described in Chapter 2 is a nice example where simple, well-known linear algebra can be highly influential when used correctly in the appropriate context. The situation might be frustrating at times, when trying to

---

* The codes can be downloaded here: https://github.com/arielamir/ThinkingProbablistically

prove something without being given in advance the precise conditions for the results to hold – yet this mimics the situation encountered so often in research. Indeed, many of the original problems arose from the author's own research experience or from (often recent) research papers, and as such reflect "natural" problems rather than contrived exercises. In other cases, the problems supplement the materials of the main chapter and essentially "teach" a classic theorem (e.g., Pólya's theorem in Chapter 2) through hands-on experience and calculations (and with the proper guidance to make it manageable, albeit occasionally challenging). We made a conscious choice to make the problems less defined and avoid almost categorically problems of the form "Prove that X takes the form of Y under the assumptions Z." The philosophy behind this choice is to allow this book (and the problems) to serve as a bridge between introducing the concepts and doing research on related topics. The typical lack of such bridges is nicely articulated by Williams (2018), which was written by a graduate student based on his own first-hand experience in making the leap from undergraduate course work to graduate-level physics research. Trickier problems will be denoted by a * (hard) or ** (very hard), based on the previous experience of students tackling these problems.

1.1 Probabilistic Surprises

Randomness can lead to counter-intuitive effects and to novel phenomena

Research has shown that our intuition for probability is far from perfect. Problems associated with probability are often easy to formulate but hard to solve, as we shall see throughout this book. Many problems have an element of randomness to them (sometimes due to our imperfect knowledge of the system) and may be best examined via a probablisitic model.

As a "warmup," let us consider several elementary examples, where the naive expectation (of most people) fails, while a simple calculation gives a counterintuitive result. Some excellent examples that we will not consider – since they are perhaps too well known – include the Monty–Hall problem (http://en.wikipedia.org/wiki/MontyHallproblem), and the false-negative paradox (http://en.wikipedia.org/wiki/Falsepositiveparadox).

1.1.1 Example: Does the Fraction of a Rare Disease in a Population Remain Fixed Over Time? (Hardy–Weinberg Equilibrium)

Consider a rare disease associated with some recessive allele (i.e., the individual must have two copies of this gene, one from each parent, in order to be sick). The population consists of three genotypes: individuals with two copies of the dominant gene (who will not be sick), which we will denote by AA; those with one copy of the recessive gene, aA; and those with two copies of it aa, who will be sick. Each of the parents gives one of the alleles to the offspring, with equal probability. This is schematically described in the table below, describing the various possibilities for the offspring's genotype (and their probabilities in brackets) given the mother's and father's genotypes.

mother \ father	aa	AA	aA
aa	$aa(1)$	$aA(1)$	$aA\left(\frac{1}{2}\right), aa\left(\frac{1}{2}\right)$
AA	$aA(1)$	$AA(1)$	$AA\left(\frac{1}{2}\right), aA\left(\frac{1}{2}\right)$
aA	$aa\left(\frac{1}{2}\right), aA\left(\frac{1}{2}\right)$	$AA\left(\frac{1}{2}\right), aA\left(\frac{1}{2}\right)$	$aa\left(\frac{1}{4}\right), AA\left(\frac{1}{4}\right), aA\left(\frac{1}{2}\right)$

We shall denote the relative abundance of genotypes AA, aA, and aa by p, $2q$, and r, respectively (thus, by definition $p + 2q + r = 1$). Surprisingly, at the beginning of the twentieth century, it was not clear what controls the relative abundance of the three types: *What are the possible stable states? What are the dynamics starting from a generic initial condition?*

On surprises If you haven't seen this problem before, you might have some prior intuition or guesses as to what the results might be. For instance, it might be reasonable to expect that if a disease corresponding to a recessive gene is initially very rare in the population, then over time it should go extinct. This is, in fact, *not* the case, as we shall shortly see. In that sense, you may call the results "surprising." But perhaps a reader with better intuition would have guessed the correct result a priori, and will not find the result surprising at all – in that sense, the notion of a "surprising result" in science is, in fact, a rather unscientific concept. In retrospect, mathematical results cannot really be surprising ... Nevertheless, the scientific *process* itself is often driven by intuition and lacks the clarity of thought that is the luxury of hindsight, and for this reason scientists do often invoke the concept of a "surprising result." Moreover, this often reflects our expectations from prior null models that we are familiar with. For example, in Section 1.1.2 we will show a simple model suggesting an exponential distribution of the time intervals between subsequent buses reaching a station. Armed with this insight, we can say that the results described in Chapter 8, finding a distribution of time intervals between buses that is not only non-exponential but in fact non-monotonic, are surprising! But this again illustrates that our definition of surprising very much hinges on our prior knowledge, and perhaps a more (or less) mathematically sophisticated reader would not find the latter finding surprising. For a related paper, see also Amir, Lemeshko, and Tokieda (2016b).

Remarkably, it was not until 1908 that the mathematician G. H. Hardy sent a letter to the editor of *Science* magazine clearing up this issue (Hardy 1908). His letter became a cornerstone of genetics (known today as the Hardy–Weinberg equilibrium, a name also crediting the independent contributions of Wilhelm Weinberg). The model and calculations are extremely simple. Assuming a well-mixed population in its nth generation, let us compute the abundance of the three genotypes in the $n + 1$ generation, assuming for simplicity random mating between the three genotypes. Using the table, it is straightforward to work out that the equations relating the fractions in one generation to the next are:

$$p_{n+1} = p^2 + 2pq + q^2 = (p+q)^2. \tag{1.1}$$

$$r_{n+1} = r^2 + 2qr + q^2 = (r+q)^2. \tag{1.2}$$

$$2q_{n+1} = 2q^2 + 2pq + 2qr + 2pr = 2(p+q)(r+q). \tag{1.3}$$

Note that we dropped the n subscript on the RHS (the abbreviation we will use for "right-hand side" through the text) to make the notation less cumbersome. As a sanity check, you can check that these sum up to $(p + 2q + r)^2 = 1$.

If we reach a stationary ("equilibrium") state, then $p_{n+1} = p_n$, etc. This implies that

$$pr = q^2. \tag{1.4}$$

Hence q is the geometric mean of p and r at equilibrium. Is this a sufficient condition for equilibrium? The answer is yes, since the first equation becomes

$$p = p^2 + 2pq + pr = p(p + 2q + r) = p \tag{1.5}$$

(and the second equation has the same structure – can you see why there is no need to check the third?).

Finally, how long would it take us to reach this state starting from general initial conditions p_1, q_1, and r_1? Note that $q_{n+1} = (p+q)(r+q)$, hence:

$$q_2^2 = (p_1 + q_1)^2 (r_1 + q_1)^2 = p_2 r_2.$$

So equilibrium is precisely established after a single generation! Importantly, in sharp contrast to the biologists' prior belief, it can have an arbitrarily small – but stable – value of r. Things are different, in fact, in the subtle variant of this problem studied in Problem 1.2.

1.1.2   Example: Bus Timings, Random vs. Ordered

In bus station A, buses are regularly sent off every 6 minutes. In station B, on the other hand, the manager throws a die every minute and sends off a bus if they get a "6."

Q: How many buses leave per day? What is the average time between buses?

It is clear that on average $60 \cdot 24/6 = 240$ buses will be sent out, in both cases. Hence the average time between buses is 6 minutes in both cases (in case A, the time between buses is, of course, always 6 minutes).

Q: If a person arrives at a random time to the bus station, how many minutes do they have to wait on average?

For simplicity, let us assume that the person arrived just after a potential bus arrival event.

What should we expect? In case A, it is equally probable for the waiting time to be $1, 2 \ldots, 6$ minutes, hence the average waiting time is 3.5 minutes. Try to think about case B. Given that the total number of buses per day and the average time between buses is identical, you might expect that the average waiting time would be identical too. This is not the case, as we shall shortly show: In case B, the average time between buses is also 6 minutes, but, perhaps counterintuitively, this is also the average waiting time!

To see this, let us assume that we got to the station at a random time. The probability to wait a minute until the next one is $1/6$. The probability for a 2 minute wait is $\frac{5}{6}\frac{1}{6}$, and more generally the probability to wait n minutes, p_n, is

$$p_n = (1 - p)^{n-1} p \tag{1.6}$$

(with $p = 1/6$).

Therefore, the average waiting time is

$$\langle T \rangle = \sum p_n n = \sum_{n=1}^{\infty} n(1 - p)^{n-1} p. \tag{1.7}$$

Without the n in front, this would be a geometric series. To deal with it, define $q \equiv 1 - p$, and note that

$$\sum_{n=0}^{\infty} q^n = 1/(1 - q). \tag{1.8}$$

Taking the derivative with respect to q gives us the following relation:

$$\sum_{n=0}^{\infty} nq^{n-1} = 1/(1 - q)^2. \tag{1.9}$$

Therefore, our sum in Eq. (1.7) equals

$$\langle T \rangle = p/(1 - q)^2 = (1/6)/(1/6)^2 = 6. \tag{1.10}$$

Looking back, this makes perfect sense, since the fact that a bus just left does not "help" us regarding the next one – the process has no memory. Interestingly, this example is relevant for the physics of a (classical) model of electron transport, known as the Drude model – where our calculations imply that an additional factor of "2" should *not* be present in the final result.

What about the distribution of time gaps? It is given by Eq. (1.6), and is therefore *exponential*. This process is a simple example of a random process, and in the continuum limit where the time interval is vanishingly small this is known as a Poisson process (see Appendix A for the related *Poisson distribution*, describing the probability distribution of the *number* of events occurring within a fixed time interval).

A note on terminology Throughout this book, we will follow the physicists' terminology of referring to a probability density function (pdf) as a "probability distribution," and referring to a "cumulative distribution" for the cumulative distribution function (cdf). Moreover, for a real random variable X we will denote the probability distribution by $p(x)$, rather than the notation f_X often used in mathematics. **Further notational details are provided in Appendix F.**

What about real buses? An online blog analyzed the transportation system in London and showed that it is Poissonian (i.e., corresponds to the aforementioned random bus scheduling) (http://jasmcole.com/2015/03/02/two-come-along-at-once/). This implies that the system is not optimal (since we can get buses coming in "bunches," as well as very long waits). On the other hand, later in the book (Chapter 8) we will see a case where buses were not Poissonian but also not uniform – the distribution was very different from exponential (the Poisson case) but was not narrowly peaked (the uniform case). Interestingly, it vanished at zero separation – buses "repelled" each other. It was found to be well described by the results of random matrix theory, which we shall cover in Chapter 8.

1.1.3 Bertrand's Paradox: The Importance of Specifying the Ensemble

Consider an equilateral triangle inscribed in a circle. Suppose a chord of the circle is chosen at random. What is the probability that the chord is longer than a side of the triangle?

We shall now show three "reasonable" methods for choosing a random chord, which will give us $1/2$, $1/3$, and $1/4$ as an answer, respectively. This is known as "Bertrand's paradox" (Bertrand 1907).

Method 1: Choosing the endpoints. Let us choose the two endpoints of the chord at random. The chord is longer than the side of the triangle in $1/3$ of the cases – as is illustrated in Fig. 1.1

Method 2: Choosing the midpoint. What about if we choose a point randomly and uniformly in the circle, and define it to be the middle of the chord? From the

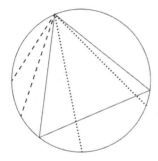

Figure 1.1 Method 1: Choosing the endpoints of the chord.

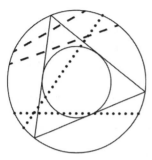

Figure 1.2 Method 2: Choosing the chord midpoint randomly and uniformly in the circle.

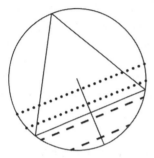

Figure 1.3 Method 3: Defining the chord midpoint by choosing a point along the radius.

construction of Fig. 1.2, we see that when the point falls within the inner circle the chord will be long enough. Its radius is $R \sin(30) = R/2$, hence its area is $1/4$ times that of the outer circle – therefore, the probability will be $1/4$.

Method 3: Choosing a point along the radius to define the midpoint. If we choose the chord midpoint along the radius of the circle with uniform probability, the chord will be long enough when the chosen point is sufficiently close to the center – it is easy to see that the triangle bisects the radius, so in this case the probability will be $1/2$ (see Fig. 1.3).

Importantly, there is no right or wrong answer – but the point is that one has to describe the way through which the "random" choice is made to fully describe the problem.

1.1.4   Benford's Law: The First Digit Distribution

Another related example where the lack of specification of the random ensemble leads to rather counterintuitive results is associated with Benford's law (named after Frank Benford, yet discovered by Simon Newcomb a few decades beforehand!). It is an empirical observation that for many "natural" datasets the distribution of the first digit is very far from uniform, see Fig. 1.4. The formula that the data is compared with is one where the relative abundance of the digit d is proportional to

$$p_d \propto \log(1 + 1/d) \tag{1.11}$$

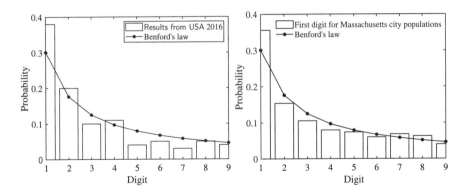

Figure 1.4 (left) Example of a dataset approximately following Benford's law, obtained by using readily available data for the vote total of the candidates in the 2016 elections in the USA across the different states (data from Wikipedia). You can easily test other datasets yourself. A similar analysis was used to suggest fraud in the 2009 Iranian elections (Battersby 2009). (right) Similar analysis of city population size in Massachusetts, based on the 2010 census data (www.togetherweteach.com/TWTIC/uscityinfo/21ma/mapopr/21mapr.htm).

(this implies that 1 occurs in about 30 % of cases and 9 in less than 5!) Although here it makes no difference, log refers to the natural logarithm throughout the text, unless otherwise specified.

Clearly, this is not always true, e.g., MATLAB's random number generator closely follows a uniform distribution, and hence the distribution of the first digit will be uniform. But it turns out to be closely followed for, e.g., tax returns, city populations, election results, physics constants, etc. What do these have in common? The random variable is broadly distributed, i.e., the distribution spans many decades, and it is far from uniform. In fact, to get Eq. (1.11), we need assume that the *logarithm* of the distribution is uniformly distributed over a large number of decades, as we shall now show. If x is such a random variable, with $y = \log(x)$, then for values of y within the support of the uniform distribution we have

$$p(x)dx = g(y)dy = Cdy,\qquad(1.12)$$

where $p(x)$ and $g(y)$ are the probability distributions, and $C = \frac{1}{y_{max}-y_{min}}$. Hence:

$$p(x) = C|dy/dx| = C/x\qquad(1.13)$$

(see Appendix A for a reminder on working with such a change of variables). Therefore, $p(x)$ rapidly increases at low values of x (until dropping to zero at values of x below the lower cutoff – set by the cutoff of the assumed uniform distribution of y).

The relative abundance of the digit 1 is therefore

$$p_1 = \int_1^2 p(x)dx + \int_{10}^{20} p(x)dx \ldots \approx C \cdot D \log(2/1),\qquad(1.14)$$

where D is the number of decades spanned by the distribution $p(x)$. Similarly

$$p_d \propto \int_d^{d+1} \frac{1}{x}dx = \log(1 + 1/d),\qquad(1.15)$$

and Benford's law follows.

The importance of specifying the ensemble Although the problems are very different in nature, there is a deep analogy between Bertrand's paradox and Benford's law in the following sense: In both cases the "surprising result" comes from a lack of definition of the random ensemble involved. In Bertrand's paradox case, it is due to our loose phrasing of "random." In the case of Benford's law, it is manifested in our misconception that given that we are looking at a random variable, the distribution of the first digit should be uniform – this would indeed be true if the random variable is drawn from a broad, *uniform* distribution, but such distributions typically do not correspond to naturally occurring datasets.

It is easy to show that Benford's law arises if we assume that the distribution of the first digit is *scale invariant* (e.g., the tax returns can be made in dollars or euros). But why do such broad distributions arise in nature so often? One argument that can be made to rationalize Benford's law relies on *multiplicative processes*, an idea that (potentially) traces back to Shockley. He noticed that the productivity of physicists at Bell labs – quantified in terms of their publication number – follows a log-normal distribution, and wanted to rationalize this observation (Shockley 1957). We repeat the argument here, albeit for the example of the population of a city, x.

This depends on a large number N of "independent" variables: the availability of water, the weather, the quality of the soil, etc. If we assume that:

$$x = \prod_{i=1}^{N} x_1 \cdot x_2 \ldots x_N,$$
(1.16)

with x_i some random, independent variables (e.g., drawn from a Gaussian distribution), then the distribution of the *logarithm* of x can be approximated by a Gaussian (by the central limit theorem, see Appendix A), hence $p(x)$ will be a log-normal distribution – which is very similar to the uniform distribution we assumed above, and Benford's law will approximately follow.

For another example where a similar argument is invoked to explain the observation of a log-normal distribution of file sizes, see Downey (2001). Another example of a variant of this argument relates to the logarithmic, slow relaxations observed in nature: see Amir, Oreg, and Imry (2012). We will revisit this example in Chapter 6 in more detail. Finally, the log-normal distribution and the multiplicative mechanism outlined above also pop up in the "numerology" context of Amir, Lemeshko, and Tokieda (2016b).

1.2   Summary

In this chapter we introduced a somewhat eclectic set of problems where probability leads to (perhaps) surprising results. Hardy–Weinberg involved genetics and population dynamics, where we derived a necessary and sufficient condition for the relation of the three possible genotypes. We discussed the important example of a Poisson process, motivated by considering a simple model for the statistics of buses arriving at a station. Finally, we exemplified the huge importance of specifying the random ensembles we are dealing with, first by following Bertrand and considering a geometric "paradox" (resulting from our incomplete definition of the problem), and next by discussing (and attempting to rationalize) the empirical observation of a rather universal (but non-uniform) distribution of the first digit associated with naturally occurring datasets.

For further reading See Mlodinow (2009) for an elementary but amusing book on surprises associated with probability and common pitfalls, with interesting historical anecdotes.

1.3   Exercises

1.1   Probability Warm-up
In a certain village, each family has children until their first daughter is born, after which they stop. What is the ratio of the expected number of boys to girls in the village? Assume a girl or boy is born with probability $1/2$.

1.2   Hardy–Weinberg Revisited
Consider a rare, *lethal* genetic disease, such that those who carry two copies of the recessive gene (aa) will not live to reproduce.

(a)   Derive the equations connecting the abundances of the AA, Aa, and aa genotypes (p, $2q$, and r) between subsequent generations.
(b)   What happens at long times?

1.3   Benford's Law and Scale Invariance
Prove that Benford's law is a consequence of scale invariance: If the first digit distribution does not depend on units (e.g., measuring income in dollars vs euros), Benford's law follows.

1.4   Benford's Law for Second Digit

(a)   Using the same assumptions behind the derivation of Benford's law, derive the probability distribution of the first two digits of each number.
(b)   What is the probability of finding 3 as the second digit of a number following Benford's law?

1.5   Drude Model of Electrical Conduction
The microscopic picture of the Drude model is that, while moving through a metal, electrons can randomly collide with ions. Interactions between electrons are neglected, and electron–ion interactions are considered short range. The probability of a single electron collision in any infinitesimal interval of time of duration dt is dt/τ (and the probability of no collision is $1 - dt/\tau$).

(a) Consider a given electron in the metal. Determine the probability of the electron not colliding within the time interval $[0, t]$.

(b) Assume that a given electron scatters at $t = 0$ and let T be the time of the following scattering event. Find the probability of the event that $t < T < t + dt$. Also calculate the expected time $\langle T \rangle$ between the two collisions.

(c) Let $t = 0$ be an arbitrary observation time. Let T_n be the time until the next collision after $t = 0$ and T_l be the time since the last collision before $t = 0$; Consider the random variable $T = T_l + T_n$. Determine the distributions of T_l and T_n, and from them deduce $\langle T \rangle$. Does your answer agree with the result in part (b)? If not, explain the discrepancy between the two answers.

1.6 Mutating Genome*

Consider an organism with a genome in which mutations happen as a Poisson process with rate U. Assume that all mutations are neutral (i.e., they do not affect the rate of reproduction). Assume the genome is large enough that the mutations always happen at different loci (this is known as the infinite sites model) and are irreversible. We start at $t = 0$ when there are no mutations.

(a) What is the probability that the genome does not obtain any new mutations within the time interval $[0, t)$?

(b) What is the expected number of mutations for time T?

(c) Consider a population following the Wright–Fisher model: At each generation, each of N individuals reproduces, but we keep the population size fixed by randomly sampling N of the $2N$ newborns. Find the probability of two individuals having their "first" (latest chronologically) common ancestor t generations ago, $P(t)$. Hint: Go backwards in time with discrete time steps. What is the continuum limit of this probability? (i.e., the result for a large population size).

(d) Let us add mutations to the Wright–Fisher model. Assume we sample two individuals that follow two different lineages for precisely t generations (i.e., their first common ancestor occured t generations ago). What is $P(\pi|t)$, the probability of π mutations arising during the t generations?

(e) What is $P(\pi)$, the probability of two individuals being separated by π mutations after they were born from the same parent? What is the expected value of π? (You may work in the continuum limit as in (c), corresponding to a large population size $N \gg 1$).

Problem credit: Jiseon Min.

2 Random Walks

These motions were such as to satisfy me ... that they arose neither from currents in the fluid, nor from its gradual evaporation, but belonged to the particle itself

(Robert Brown)

Consider a small particle suspended in a liquid. Due to the constant collisions with the surrounding liquid molecules, the path followed by the particle will be erratic, as was first noticed by Robert Brown in the nineteenth century in experiments where he was tracking the motion of pollen grains (Brown 1828). As a result, this is often known as Brownian motion – see also Pearle *et al.* (2010) for a modern take on Brown's experiments, and Fig. 2.1 for a later example of a quantitative study of particle diffusion by Jean Baptiste Perrin, which we will mention again in Chapter 3. This process was modeled by Albert Einstein, who derived the so-called diffusion equation, and understood the nature of the particle's dynamics. In this chapter, we will first study a simplified model for diffusion, where, following Einstein's original derivation from 1905, time will be discrete (i.e., at every time step the particle will move in some random direction). We will understand how far the particle typically gets after making N moves, and what its probability distribution is. These ideas are central in understanding numerous processes around us: from the dynamics of diffusing particles in liquids as well as in living cells, to modeling the dynamics of the stock market (which we will get to later in the book, in Chapter 3). In fact, the concept of "random walks," as this dynamics is often referred to, will also play an important role in our discussion of "Google PageRank," the algorithm at the heart of the search engine.

Asking the right question In science, it is often as important (and hard) to ask the right question than to come up with the right answer. The great statistician Karl Pearson (1905) sent a letter to *Nature* magazine posing, quite literally, the problem of the random walker: given that at every step a person chooses a random direction and walks a constant number of steps, what is the distribution of their position after N steps? It is remarkable that random walks were only introduced in the twentieth century, and a strange coincidence that they were almost simultaneously suggested by Pearson and Einstein.

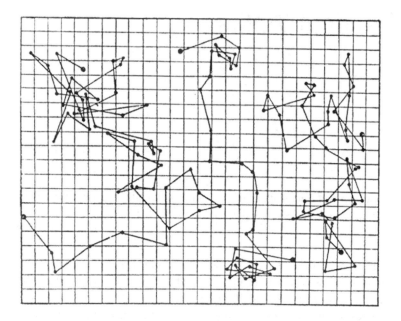

Figure 2.1 An experimental observation by Perrin of the diffusion of small (spherical) particles suspended in a liquid. From Perrin (2013).

2.1   Random Walks in 1D

Consider a "random walker" on a one-dimensional lattice. Time will be discrete, and at every time step the walker will take a step in one of the possible two directions, with equal probability. What is the mean value of, say, coordinate x, after N steps? From symmetry, it is clearly zero. But we certainly don't expect the walker to be precisely at the origin after a large number of steps. Its typical magnitude away from the origin may be found by calculating the RMS (root-mean-square) of the trajectories, i.e., the second moment of the distance from the origin.

Let's say the walker was at position x after N steps. After an additional step, it could be in one of two possible positions. Averaging the squared distance between these two possibilities, conditioned on being at position x at time N, gives us

$$\frac{1}{2}[(x + 1)^2 + (x - 1)^2] = x^2 + 1. \tag{2.1}$$

Upon averaging over the value of x^2 at time N, we find that

$$\langle x_{N+1}^2 \rangle = \langle x_N^2 \rangle + 1; \tag{2.2}$$

therefore, repeating the argument $N - 1$ times, we find that

$$\langle x_N^2 \rangle = N. \tag{2.3}$$

This implies that the typical distance from the origin scales like $\sqrt{N}$. What about the position *distribution*? If N is even, then it is clear that the probability to be a distance M from the origin after N steps is zero for odd M, and for even M it is

$$p_M = \frac{1}{2^N}\binom{N}{R} = \frac{1}{2^N}\frac{N!}{R!\,(N-R)!},\qquad(2.4)$$

where R is the number of steps to the right, thus $R-(N-R) = 2R-N = M$. We can now evaluate this probability for $N \gg M$ using Stirling's formula, which provides an (excellent) approximation for $N!$, namely $N! \approx \sqrt{2\pi N}\left(\frac{N}{e}\right)^N$. This leads to

$$p_M \approx \frac{1}{\sqrt{2\pi}}\frac{1}{2^N}\frac{N^{N+1/2}}{R^{R+1/2}(N-R)^{N-R+1/2}}$$

$$= \frac{1}{\sqrt{2\pi}}e^{-N\log(2)+(N+1/2)\log(N)-(R+1/2)\log(R)-(N-R+1/2)\log(N-R)}.\qquad(2.5)$$

We can proceed by using our assumption that $N \gg M$, implying that R is approximately equal to $N/2$.

Choosing the appropriate limit In physics and mathematics, it is often as important to analyze the results in the relevant limit as it is to derive the exact results. Based on the previous results we derived, namely, that the variance of the random walker increases linearly in time, it seems plausible that the limit $N \gg M$ will typically hold (i.e., for a large N the probability of M not obeying this condition will be small) – can you see why?

Under this approximation we find that

$$\log(R) = \log\left(\frac{N+M}{2}\right) = \log(N/2(1+M/N)) \approx \log(N/2) + M/N - \frac{1}{2}(M/N)^2,$$

$$(2.6)$$

where we used the Taylor series expansion $\log(1+x) \approx x - x^2/2 + x^3/3\ldots$, valid for small x. Similarly, we have

$$\log(N-R) = \log\left(\frac{N-M}{2}\right) = \log(N/2(1-M/N))$$

$$\approx \log(N/2) - M/N - \frac{1}{2}(M/N)^2.\qquad(2.7)$$

Putting this together we find that

$$(R+1/2)\log(R) + (N-R+1/2)\log(N-R)$$

$$\approx (R+1/2)\left[\log(N/2) + M/N - \frac{1}{2}(M/N)^2\right]$$

$$+ (N-R+1/2)\left[\log(N/2) - M/N - \frac{1}{2}(M/N)^2\right].\qquad(2.8)$$

Simplifying leads to

$$(R + 1/2)\log(R) + (N - R + 1/2)\log(N - R) \approx (N + 1)\log(N/2) + M^2/2N.$$
$$(2.9)$$

Finally,

$$p_M \approx \frac{2}{\sqrt{2\pi N}} e^{-\frac{M^2}{2N}}.$$
$$(2.10)$$

Hence the distribution is approximately Gaussian. In fact, this had to be the case since x is a sum of independent random variables, hence according to the central limit theorem (see Appendix B for a reminder) it should indeed converge to a Gaussian distribution! Note that the distribution indeed sums up to 1, since the support of this Gaussian is only on even sites.

In summary, we learnt that

1. A diffusing particle would get to a distance $\sim \sqrt{t}$ from its starting position after a time t.
2. The probability distribution describing the particle's position is approximately Gaussian.

Random walks in 1D are strange and interesting! The problem set at the end of the chapter will expose you to some of the peculiarities associated with random walks in one dimension. Problems 2.6 and 2.7 deal with recurrent vs. transient random walks. It turns out that in one- and two-dimensional random walks a walk will *always* return to the origin sometime in the future ("recurrent random walks") – but that this is *not* the case for higher dimensions, where the random walk is called "transient." This is known as Pólya's theorem. Given that 1D random walks are recurrent, we may ask what the "first return time" distribution is – the distribution of the time to return to the origin for the first time. Solving Problems 2.1, 2.3 or 2.9 will show you that this is a power-law distribution, which, remarkably, has diverging mean – so while you *always* return to the origin, the mean time to return is *infinite*! This property is discussed in detail in Krapivsky, Redner, and Ben-Naim (2010). In fact, mean first passage times of random walkers (in any dimension) have a beautiful analogy with resistor networks (see also Doyle and Snell 1984) and relations with harmonic functions. Additional peculiarities arise if we consider the distribution of the *last* time a random walker returns to the origin within a given time interval. This is studied in Problem 2.2 (see also Kostinski and Amir 2016).

2.2   Derivation of the Diffusion Equation for Random Walks in Arbitrary Spatial Dimension

We shall now approach the problem with more generality, following nearly precisely the derivation by Einstein. We will work in 3D but the approach would be the same in any dimension. The approach will have discrete time steps, but the step direction will be a continuous random variable, described by a probability distribution $g(\vec{\Delta})$ (here $\vec{\Delta}$ is a vector describing the step in the 3D space – not to be confused with the Laplacian operator!). We will not limit the random walker to a lattice, though it is possible to implement such a scenario by taking $g(\vec{\Delta})$ to be a sum of δ-functions (can you see how?).

We will seek to obtain the probability distribution $p(\vec{r})$, i.e., $p(\vec{r})dV$ will be the probability to find the particle in a volume dV around the point $\vec{r}$ (at some point in time corresponding to a a given number of steps). If the original problem is cast on a lattice, this distribution will be relevant to describe the *coarse grained* problem, when we shall zoom-out far enough such that we will not care about the details at the level of the lattice constant.

If at time t the probability distribution is described by $p(\vec{r},t)$, let us consider what it will be a time τ later, where τ denotes the duration of each step. As you can guess, in a realistic scenario the time of a step is non-constant, and τ would be the mean step time. Thus, we haven't lost too much in making time discrete – but we did make an assumption that a mean time *exists*. In Chapter 7 we will revisit this point, and show that in certain situations when the mean time diverges, we can get *subdiffusion* (slower spread of the probability distribution over time compared with diffusion).

To find $p(\vec{r},t+\tau)$, we need to integrate over all space, and consider the probability to have the "right" jump size to bring us to $\vec{r}$. This leads to

$$p(\vec{r},t+\tau) = \int p(\vec{r}-\vec{\Delta},t)d^3\vec{\Delta}g(\vec{\Delta}). \tag{2.11}$$

To proceed, we will Taylor expand p, assuming that the probability distribution is smooth on the scale of the typical jump. If we expand it to first order, we will find that

$$p(\vec{r}-\vec{\Delta}) \approx p(\vec{r}) - (\nabla p)\cdot\vec{\Delta}. \tag{2.12}$$

If the diffusion process is isotropic in space, there is no difference between a jump in the $\pm\vec{\Delta}$ direction, so the integral associated with the second term trivially vanishes:

$$\int ((\nabla p)\cdot\vec{\Delta})g(\vec{\Delta})d^3\vec{\Delta} = 0. \tag{2.13}$$

This means we have to expand to second order:

$$p(\vec{r}-\vec{\Delta}) \approx p(\vec{r}) - (\nabla p)\cdot\vec{\Delta} + \frac{1}{2}\sum_{i,j}\frac{\partial^2 p}{\partial x_i \partial x_j}\bigg|_{\vec{r}}\Delta_i\Delta_j. \tag{2.14}$$

To which order should we expand? In deriving both Eqs. 2.6 and 2.14 we had to decide to which order we should Taylor expand a function. Were we to only retain the first-order term in the expansion, the results would have been nonsensical. In principle, we should make sure that the next order in the expansion is negligible compared with the terms we have kept, but we will often omit this step and rely on our physical intuition instead in deciding the "correct" order of expansion.

Plugging this into Eq. (2.11), we find that

$$p(\vec{r}, t + \tau) - p(\vec{r}, t) \approx \int \frac{1}{2} \sum_{i,j} \frac{\partial^2 p}{\partial x_i \partial x_j} \Delta_i \Delta_j g(\vec{\Delta}) d^3 \vec{\Delta}. \tag{2.15}$$

Once again we may make further progress if we make assumptions regarding the symmetries associated with g: If we assume isotropic diffusion then $g(x, y, z) = g(-x, y, z)$, etc., implying that the only terms that would survive in the integration are the "diagonal" ones, and they would all be equal. Hence

$$p(\vec{r}, t + \tau) - p(\vec{r}, t) \approx \int \frac{1}{2} \sum_i \frac{\partial^2 p}{\partial x_i^2} \Delta_i^2 g(\vec{\Delta}) d^3 \vec{\Delta}. \tag{2.16}$$

From symmetry, we can replace $\int \Delta_i^2 g(\vec{\Delta}) d^3 \vec{\Delta} = \frac{1}{3} \int ||\vec{\Delta}||^2 g(\vec{\Delta}) d^3 \vec{\Delta} = \frac{1}{3} \langle \Delta^2 \rangle$, where $\langle \Delta^2 \rangle$ is the second moment of the jump length distribution. Therefore, we find that

$$p(\vec{r}, t + \tau) - p(\vec{r}, t) \approx \frac{1}{6} \nabla^2 p \langle \Delta^2 \rangle. \tag{2.17}$$

Replacing $p(\vec{r}, t + \tau) - p(\vec{r}, t) \approx \frac{\partial p}{\partial t} \tau$, we finally reach the famous diffusion equation

$$\frac{\partial p}{\partial t} = D \nabla^2 p, \tag{2.18}$$

with $D \equiv \frac{1}{6\tau} \langle \Delta^2 \rangle$.

In the case of a random walker in dimension d, repeating the same derivation will reproduce Eq. (2.18) albeit with a diffusion constant $D \equiv \frac{1}{2d\tau} \langle \Delta^2 \rangle$.

To which order should we expand? Again! One might argue, correctly, that to better approximate $p(\vec{r}, t + \tau) - p(\vec{r}, t)$ we should evaluate the partial derivative with respect to time in the middle of the interval $(t, t + \tau)$. Can you see why in the above derivation it suffices to evaluate it at time t?

Notice that on the way we made another "hidden" assumption: that the second moment of the jump distribution *exists*. If it doesn't, the $\sqrt{t}$ scaling that we are familiar with will break down, and this time we will get *superdiffusion* (faster spread of the probability distribution over time compared with diffusion) – this scenario is known as a Lévy-flight. An interesting case arises when the variance of the step size diverges as well as the mean time between steps. Should we get sub or super diffusion in this case? As one may anticipate, the two effects compete with each other, and both options can occur, depending on the details. We will study this in detail later on in the book, in Chapter 7.

Returning to Eq. (2.18), we will now find the probability distribution as a function of space and time for a particle found at the origin at time $t = 0$. To proceed, let us Fourier transform the equation, denoting by $\hat{p}$ the F.T. of the distribution (with respect to space), to find that

$$\frac{\partial \hat{p}}{\partial t} = -Dk^2 \hat{p} \tag{2.19}$$

(see Appendix C for a refresher on Fourier transforms). Here we used the fact that the F.T. of a derivative gives $i \cdot k$:

$$\hat{f}' \equiv \int_{-\infty}^{\infty} \frac{df(x)}{dx} e^{-ik_x \cdot x} dx = fe^{-ik_x \cdot x}|_{-\infty}^{\infty} + ik_x \int_{-\infty}^{\infty} fe^{-ik_x \cdot x} dx = ik_x \hat{f}. \tag{2.20}$$

Note: the F.T. is sometimes defined with an additional $\frac{1}{\sqrt{2\pi}}$ – here this would not have made any difference, but later on in the book we will actually have to be quite mindful of these subtleties.

In this way, we decoupled the different Fourier components, and clearly each of them will decay exponentially, with a rate Dk^2: The long wavelength modes, corresponding to small k, will have diverging timescales. What is the intuition for this? Consider an initial condition where the probability distribution is sinusoidal, with a wavelength λ, as illustrated in Fig. 2.2. Eventually, particles will diffuse around to give a uniform probability distribution, but to do that, they have to cross a characteristic length λ. How long does this take? From the previous analysis we know that $\lambda \sim \sqrt{Dt}$, hence this time will scale as λ^2/D, consistent with the above expression for the decay rate constant of this Fourier mode.

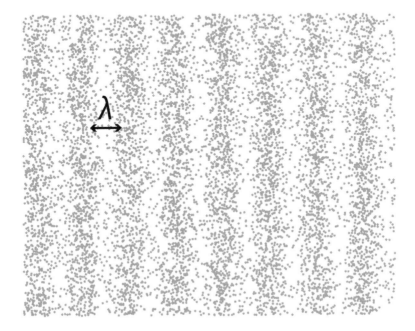

Figure 2.2 An initial configuration in which the density is modulated sinusoidally with a wavelength λ.

Fourier analysis It is appropriate that we are using Fourier transforms to solve the diffusion equation – back in 1822, Joseph Fourier invented the notion of a Fourier series in order to solve the heat equation – the equation governing the flow of heat in a solid – mathematically identical to the diffusion equation we are studying here.

Next, we assume that our initial condition (at time $t = 0$) has a probability distribution concentrated around the origin (i.e., a δ-function at the origin), and we ask what the probability distribution is at some later time t. In fact, the solution of this problem is known as *Green's function*, and knowledge of it will allow us to cast the solution of *any* initial condition in terms of a convolution with Green's function – see Problem 2.4. Since the F.T. of a δ-function at the origin is a constant (i.e., all Fourier modes are equally excited) we have

$$\hat{p}(\vec{k}, t = 0) = 1. \tag{2.21}$$

According to Eq. (2.19), we have

$$\hat{p}(\vec{k}, t) = e^{-Dk^2 t}. \tag{2.22}$$

Scaling approach to solving the equation In Problem 2.11 a different route to solution of the diffusion equation in 1D is presented, which relies on guessing a scaling solution of a particular form. This is a commonly used method to solve PDE's of this sort, and is useful also for nonlinear PDE's where the Fourier analysis approach presented above fails.

The last step would be to inverse-Fourier this function to find what the probability distribution looks like in real space:

$$p(\vec{r}, t) = \frac{1}{(2\pi)^3} \int e^{-Dk^2 t} e^{i(k_x \cdot x + k_y \cdot y + k_z \cdot z)} dk_x dk_y dk_z. \tag{2.23}$$

Clearly, this separates into a product of three integrals (which means we can actually know the answer at this stage without further calculations if we draw on the 1D case!). Let us evaluate one of them:

$$\frac{1}{2\pi} \int_{-\infty}^{\infty} e^{-Dk_x^2 t} e^{ik_x \cdot x} dk_x = ?. \tag{2.24}$$

To proceed, we will use a trick that is often used in the context of Gaussian integrals, "completing the square":

$$I = \frac{1}{2\pi} \int_{-\infty}^{\infty} e^{-Dt(k_x - \frac{ix}{2Dt})^2} e^{-\frac{x^2}{4Dt}} dk_x. \tag{2.25}$$

The last term can be taken out of the integral, since it does not depend on k_x, hence

$$I = \frac{1}{2\pi} e^{-\frac{x^2}{4Dt}} \int_{-\infty}^{\infty} e^{-Dt(k_x - \frac{ix}{2Dt})^2} dk_x. \tag{2.26}$$

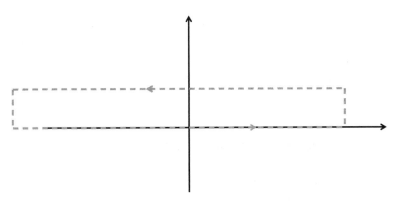

Figure 2.3 Consider the contour integral of the analytic function $f(z) = e^{-Az^2}$.

To evaluate the integral over the first term, it is tempting to make a change of variables and get rid of the $\frac{ix}{2Dt}$ term. This is incorrect, of course, and more care has to be taken when dealing with complex variables (see Appendix C for a brief summary of contour integrals and the residue theorem). It is instructive to consider a contour integral in the complex plane, as is shown in Fig. 2.3. Since the function is analytic and the contributions of the vertical parts of the contour are negligible, we find that

$$\int_{-\infty}^{\infty} e^{-Dt(k_x - \frac{ix}{2Dt})^2} dk_x = \int_{-\infty}^{\infty} e^{-Dtk_x^2} dk_x = \sqrt{\frac{\pi}{Dt}} \qquad (2.27)$$

(see Appendix A.1.2 for a detailed calculation of this one-dimensional Gaussian integral).

Putting everything together we conclude that

$$p(\vec{r},t) = \frac{1}{(\sqrt{4\pi Dt})^3} e^{-\frac{r^2}{4Dt}} \qquad (2.28)$$

(as a sanity check, you can verify that the probability distribution is normalized).

Therefore, the probability distribution is Gaussian, and its variance is

$$\int r^2 p(\vec{r},t) d^3 r = \int 3x^2 p(\vec{r},t) d^3 r = 3 \int_{-\infty}^{\infty} x^2 \frac{1}{\sqrt{4\pi Dt}} e^{-\frac{x^2}{4Dt}} dx. \qquad (2.29)$$

This can be evaluated (how?) to give

$$\langle r^2 \rangle = 6Dt. \qquad (2.30)$$

Random walks are useful Einstein's derivation was geared toward understanding diffusion of particles, but random walks have found applications in many other fields of science. Problem 2.10 explores the application of random walks to polymer physics. The next section will show why studying random walks on *networks* can lead to a useful algorithm that formed the beginning of a little search engine called Google.

In future chapters, we will also see how building on the ideas of random walks in the context of economics will lead to the celebrated Black–Scholes equation (in fact, a few years earlier than Einstein the ideas of random walks were introduced by Louis Bachelier (1900) in the context of finance).

2.3   Markov Processes and Markov Chains

Let us approach our analysis of the random walker from a new perspective, which will be easier to generalize to other networks. Within our model for diffusion on a 1D lattice, the probability to go to site j does not depend on the history of the random walker, but only on its current state – this is an example of a *Markov process* (named after Andrey Markov). A familiar childhood example of a Markov process is the game Chutes and Ladders (as well as many other board games – see also Problem 2.15). But we emphasize that whether a process is Markovian depends on the space: Consider, for example, a game where we throw a die at each round, and keep track of the running sum. This is clearly a Markov process – knowing the current sum determines the probabilities to go to the next states. But what about a game in which we reset the sum to zero each time we get two identical numbers in a row? In this case, the process is memory dependent, so it is non-Markovian. But, if we work in the space where a state is defined by a *pair*, the running sum, and the result of the last throw – then the process becomes Markovian again. It is also worth noting that in cases where time steps are discrete, the process is referred to as a *Markov chain*. In what follows we will deal with Markov chains, though in the next chapter we will study Markov processes with *continuous* time. For an extended discussion of Markov chains, see Feller (2008).

Let us denote by P the matrix describing the transition probabilities, i.e., P_{ij} will be the probability to go from i to j. The key insight to note is that for a Markov chain if we know the current probabilities to be at every site, which we will denote by the vector $\vec{p}$, and we know the matrix P, we can easily find the vector of probabilities to be at every site after an additional move. By the definition of the matrix P, the probability to be in the ith site after an additional move is given by

$$p_i^{n+1} = \sum_j p_j^n P_{ji}. \tag{2.31}$$

This can be written in more concise notation using the matrix formulation

$$\vec{p}_{n+1} = P^T \vec{p}_n, \tag{2.32}$$

where ^{T} denotes the transpose operation, and $\vec{p}_n$ a column vector the entries of which are the probabilities to be at a given site after n steps.

If we start at, say, the first site, then our initial vector of probabilities $\vec{p}_0$ will be

$$\vec{p}_0 = \begin{pmatrix} 1 \\ 0 \\ \vdots \\ 0 \end{pmatrix}, \tag{2.33}$$

and if we would like to find it after N steps, we simply have to operate with the matrix P^T on this vector N times. In other words:

$$\vec{p}_N = [P^T]^N \vec{p}_0. \tag{2.34}$$

This operation is computationally cheap, hence this is a much more efficient way of getting the answer than the simulation of many random walks and averaging over them.

2.4   Google PageRank: Random Walks on Networks as an Example of a Useful Markov Chain

We shall now describe a beautiful application of the concepts of random walks and Markov chains that contributed to one of the most influential algorithms of all time, as first described by Larry Page *et al.* (1999) in a seminal paper (which was the basis for Google).

The internet is a complex, directed graph. Imagine building a search engine, and attempting to respond to a particular search query. We would first have to find the relevant webpages (by searching the words in the text), and then choose the most appropriate one. It turns out that prior to 1998 the latter step was the more challenging, and that was the novelty which Google brought.

The idea behind PageRank is simple: We will model "surfing" on the WWW as a random walk process. We will assume that when a surfer reaches some webpage, with probability d (the "damping factor") he or she will continue to one of the pages that the current one links to, with equal probability. In practice, $d = 0.85$ is apparently used. With probability $1 - d$, the surfer is assumed to go to an arbitrary page in the web, with equal probability for all pages. The idea now is to associate the "merit" of a webpage with the (relative) number of times (i.e., frequency) that a surfer will visit the webpage starting from an arbitrary page. We will define the PageRank as the probability of being at a particular webpage in the limit of spending an infinite time surfing. Such a process only makes sense if an identical stationary distribution is reached starting from all initial conditions – which we will show is indeed the case. In other words, if the random walker surfs the web for a very long time, we will show that the probability to be in each page will approach a constant value (different for each page, usually), and that will be the "PageRank." Finally, if a page has no outgoing links, the random walker will choose a page at random from all pages, which is mathematically equivalent to having that page link to *all* other pages (including itself).

Figure 2.4 A simple network, with only two webpages.

Validating PageRank It is interesting to think how Page *et al.* knew that their newly developed search engine was indeed superior – how can that be done if there is no "benchmark" to compare against? The curious reader can look at their original paper for their "solution" to this.

2.4.1   Simple Example of the Dynamics of a Random Walker on a Network

Let's first illustrate how PageRank works on a simple example. Consider the page architecture shown in Fig. 2.4. (Note that it is also important to specify whether the webpages link to themselves or not.) As before, we denote by P the matrix describing the transition probabilities (i.e., P_{ij} will be the probability to go from i to j). For a damping factor d the matrix will be

$$P = (1-d) \begin{pmatrix} 1/2 & 1/2 \\ 1/2 & 1/2 \end{pmatrix} + d \begin{pmatrix} 1/2 & 1/2 \\ 0 & 1 \end{pmatrix} = \begin{pmatrix} 1/2 & 1/2 \\ \frac{1-d}{2} & \frac{1+d}{2} \end{pmatrix}. \quad (2.35)$$

Assuming that we start in page (1), what is the probability to be in page (1) after N steps? What happens if we ask the same question starting with page (2)?

As Eq. (2.34) shows, we have to take P^t to the Nth power.

This can be done more economically by finding the eigenvalues and eigenvectors of the matrix (see Appendix B if a refresher is needed).

Its characteristic polynomial is

$$|\lambda I - P| = (\lambda - 1/2)(\lambda - (1+d)/2) - \frac{1-d}{4}. \quad (2.36)$$

The eigenvalues are the roots of this polynomial. It is easy to check that $\lambda = 1$ is a root – we will shortly see why this has to be the case in general. Since the trace of the matrix is equal to the sum of eigenvalues, the second one must be such that they sum up to $1 + d/2$, hence it is $d/2$. We can now write the initial probability vector as a sum of the two eigenmodes, i.e.,

$$\vec{v} = a_1 \vec{v}_1 + a_2 \vec{v}_2. \quad (2.37)$$

The advantage of working with the eigenmodes is that after operating with P^T we will get $\lambda_j \vec{v}_j$, hence after operating with $(P^T)^N$ we will get $\lambda_j^N \vec{v}_j$. Therefore, the probability to be at (1) after N operations has the form:

$$p_1^N = a_1 \lambda_1^N (\vec{v}_1)_1 + a_2 \lambda_2^N (\vec{v}_2)_1. \tag{2.38}$$

We know that $\lambda_1 = 1$ and $\lambda_2 = d/2$, hence

$$p_1^N = c_1 + c_2(d/2)^N. \tag{2.39}$$

It remains to find c_1 and c_2. One way to do it would be to find the eigenmodes, and then find a_1 and a_2. But a shortcut would be to use the fact that $p_1^0 = 1$ and $p_1^1 = 1/2$, hence $c_1 + c_2 = 1; c_1 + c_2(d/2) = 1/2 \Rightarrow c_1 = \frac{1-d}{2-d}, c_2 = \frac{1}{2-d}$.

Note that, in this example, the probability to be at (1) decays exponentially to its asymptotic value $\frac{1-d}{2-d}$, with a characteristic timescale which is associated with the second eigenvalue. In general, this exponential decay can be oscillatory – corresponding to complex eigenvalues (the initial matrix is typically non-Hermitian, so the eigenvalues are not guaranteed to be real – see Appendix B for a concise summary of these linear algebra facts). As a sanity check, note that for $d = 1$ the asymptotic value to be at (1) vanishes, while for $d = 0$ it is 1/2.

Repeating the same analysis when starting with page (2), Eq. (2.39) is still intact, but now $c_1 + c_2(d/2) = \frac{1+d}{2}$, leading to $c_1 = \frac{1}{2-d}, c_2 = \frac{1-d}{2-d}$. Thus, at long times the probability to be at (2) would approach $\frac{1}{2-d}$, and the probability for (1) would approach $\frac{1-d}{2-d}$ – just as they have for the other initial condition. We will now generalize this example for arbitrary networks.

2.4.2 Existence and Uniqueness of the Stationary Distribution in the Context of PageRank

It is natural to ask whether a stationary distribution exists – one that is invariant under the operation of the matrix P^T – and whether we always converge to it starting from any initial conditions. We shall see that for the PageRank setup with damping factor $d < 1$ such a distribution indeed exists and we necessarily converge to it. For $d = 1$ we can construct examples where the probability vector would never converge to a constant one. For example, consider a network with two pages, and

$$P = \begin{pmatrix} 0 & 1 \\ 1 & 0 \end{pmatrix}. \tag{2.40}$$

If we start in page (1), then the probability to be at (1) will oscillate forever between the values of 0 and 1, and hence does not converge to a constant. This non-convergent behavior will not occur once we put in the "damping factor," as we shall shortly prove.

First, let us prove that there is an eigenvector with eigenvalue of precisely 1. For the matrix P, this is easy to see, by considering the product of this matrix by a vector $\vec{v}$ the entries of which are all equal to 1. We have

$$\sum_j P_{ij}v_j = \sum_j P_{ij} = 1 = v_i. \tag{2.41}$$

Therefore, there must also be an eigenvalue 1 for the matrix P^T (since they share the same characteristic polynomial). The corresponding eigenvector will therefore not decay in time, and is a proof for the existence of a stationary distribution. What about its uniqueness? To establish that, we will show that all other eigenvalues will have magnitude smaller than 1. Consider an eigenvector with eigenvalue λ and entries v_j. Let us assume that the entry with the largest magnitude is the mth one. Without loss of generality, we can choose it to be real and positive (can you see why?). Consider now the matrix P operating on the vector, and in particular the summation corresponding to the mth element. It reads:

$$\sum_j P_{mj}v_j = \lambda v_m. \tag{2.42}$$

Additionally, we have $\sum_j P_{mj} = 1$. Considering that the entries of P are positive, this implies that

$$\left| \sum_j P_{mj}v_j \right| \le v_m \sum_j P_{mj} = v_m. \tag{2.43}$$

Hence all eigenvalues have a magnitude smaller than or equal to 1. When can the equality occur? In order for it to occur, we need every v_j that multiplies a non-vanishing matrix element P_{mj} to be equal to v_m. But for the PageRank matrix, for $d < 1$, all matrix entries are nonzero, hence *all* the elements of $\vec{v}$ have to be equal – proving that the only eigenvalue with magnitude 1 is the eigenvector we guessed previously. Think about the oscillatory example shown above, and explain why the eigenvalue -1 of that matrix does not violate this proof! Note that while this proof was for the eigenvalues of P, the same holds for its transpose (since they share the same characteristic polynomial). Also, note that for symmetric matrices the unique eigenmode the entries of which are equal will be proportional (up to trivial normalization) to the stationary distribution. The fact that the matrix corresponding to the WWW is asymmetric is essential for this algorithm to be useful!

What we just proved is part of the Perron–Frobenius theorem (the full theorem being slightly more general (Feller 2008)).

2.4.3   Explicit Formula for PageRank

By now we understand that a unique eigenvector with eigenvalue 1 exists, and understand why it is a potentially good way of ranking the web. For the case of PageRank we can write an explicit formula for it. Consider the product of P^T on the PageRank vector. This, as we now know, results in precisely the PageRank vector. We also note that $P^T_{A,B}$ corresponds to the transition rate from page B to page A, which for the PageRank algorithm equals $\frac{1-d}{N}$ if the page *doesn't* link to A, and $\frac{1-d}{N} + \frac{d}{deg(B)}$ if it does, where $deg(B)$ is the *outgoing* degree of the node B (i.e., the number of pages that it links to). Calculating explicitly the Ath entry of the product of P^T and the PageRank vector then leads to

$$PR(A) = \frac{1-d}{N} + d \sum_{B \to A} \frac{PR(B)}{deg(B)}, \qquad (2.44)$$

where the sum is over pages which have a link to A.

Another way of thinking about it You can view Eq. (2.44) as basically saying that the importance of a page is the weighted sum of the importance of all the pages that link to it. How important are *they*? The equation applies in the same way to all pages, and that's why we get an eigenvalue problem.

In fact, there are many other contexts in which *ranking* a set of objects (not web-pages) is useful. Perhaps not surprisingly given its simplicity, the PageRank algorithm, although most prominently utilized by Page *et al.* in the context of ranking webpages, was actually used earlier in various contexts. For example, Pinski and Narin (1976) used it in the context of ranking scientific journals and Keener (1993) in the context of ranking football teams. See also Ermann, Frahm, and Shepelyansky (2015) for various additional applications and an extended discussion of the algorithm.

2.4.4   Convergence to Stationary Distribution Starting from Arbitrary Initial Conditions

So far we have proved the uniqueness and existence of a stationary distribution, and saw one example where starting from some initial conditions we converge to it. But how do we know that this is true in general? (For example, perhaps the system can undergo oscillations between different states.)

Typically, the transition matrix P will be diagonalizable: It will be similar to a diagonal matrix, which has the eigenvalues on the diagonal. In the basis in which P is diagonal, taking the matrix to a large power would imply that all eigenvalues will decay to 0 (since their magnitude is smaller than 1) except for the one corresponding to the stationary distribution: Hence we will converge to it, no matter what the initial conditions are.

The situation is trickier in cases where the matrix is not diagonalizable (can you think of one that obeys our assumptions?) In that case, although the matrix is *not* similar to a diagonal matrix, it can be shown that it is similar to an *upper triangular* matrix (for example, by using the Schur decomposition or the Jordan form of the matrix). This matrix still has the N eigenvalues on the diagonal. It is then possible to show that taking that matrix to a large power leads to a matrix where all but one row vanish – implying the convergence to the stationary distribution; Can you see why? (see also Problem 2.14).

2.5   Relation between Markov Chains and the Diffusion Equation

Let us now relate the PageRank formulation to the diffusion equation which we derived before, focusing on the 1D case for concreteness. It will also be convenient to work with periodic boundary conditions – so you should envision a set of N sites organized at regular intervals on a circle, with the random walker jumping clockwise or anticlockwise at every time step. The matrix corresponding to this network for $N = 5$ is

$$P = \begin{pmatrix} 0 & 1/2 & 0 & 0 & 1/2 \\ 1/2 & 0 & 1/2 & 0 & 0 \\ 0 & 1/2 & 0 & 1/2 & 0 \\ 0 & 0 & 1/2 & 0 & 1/2 \\ 1/2 & 0 & 0 & 1/2 & 0 \end{pmatrix}. \tag{2.45}$$

In the following we will prove that for odd N there will be convergence to a unique stationary distribution for any initial conditions (note that the matrix does not satisfy the PageRank conditions, so this is not obvious a priori). Since the matrix is symmetric, the stationary distribution is uniform on all sites. But how long will it take the stationary distribution to be established, if we start at one of the sites? According to the previous analysis in terms of the eigenmodes, this time is set by the eigenvalue with the slowest decay rate, which is the one with the second to largest real part (since the stationary distribution corresponds to that with the largest real part, namely 1). In fact, this matrix is real and symmetric so all eigenvalues are real. Moreover, its particular structure allows for an exact solution of the eigenvalues and eigenmodes.

Let us "guess" a solution in the form of plane-waves, i.e.,

$$A_d = e^{ikd}. \tag{2.46}$$

Multiplying this vector by the matrix P, we find the following equation for every entry c save for the first and the last:

$$\sum_d P_{cd} A_d = \sum_d P_{cd} e^{ikd} = \frac{1}{2}\left(e^{ik(c-1)} + e^{ik(c+1)}\right) = \lambda A_c = \lambda e^{ikc}. \tag{2.47}$$

This would indeed be a solution, as long as

$$\lambda = \cos(k). \tag{2.48}$$

It remains to consider the equations for the first and last entries. It is easy to see that they would also be satisfied if and only if

$$e^{ikN} = 1. \tag{2.49}$$

This implies that the allowed values for k are: $k = 2\pi M/N$, with M an integer. It is easy to check that taking $M = 0, 1..N - 1$ leads to N eigenvectors which are all *orthogonal* to each other, and thus are independent and span the vector space.

Relation to Fast Fourier Transform It is interesting to point out that the eigenmodes of this particular matrix play a pivotal role in a remarkable algorithm known as "Fast Fourier Transform" (FFT), which allows one, among other applications, to multiply two N digit number using $O(N \log N)$ operations rather than the naive N^2 you might expect (Knuth 1998).

Since the eigenvalues are equal to the cosine of k, it is clear that other than the unique $\lambda = 1$, all other eigenvalues are doubly degenerate (save the eigenvalue -1 for even N). The (doubly degenerate) eigenvalue closest to 1 (but distinct from it) corresponds to $M = 1$ or $M = N - 1$, and its value is

$$\lambda = \cos(2\pi/N) \approx 1 - \frac{1}{2}(2\pi/N)^2. \tag{2.50}$$

Therefore, after t steps the amplitude corresponding to this eigenmode would decay as

$$\lambda^t = \left(1 - \frac{1}{2}(2\pi/N)^2\right)^t. \tag{2.51}$$

We can now use the relation $(1 - x/t)^t \approx e^{-x}$, to approximate

$$\lambda^t \approx e^{-\frac{t}{2}(2\pi/N)^2}. \tag{2.52}$$

In fact, if one wants to quantify the correction to this approximation, it is helpful to write

$$(1 - x/t)^t = e^{t \log(1-x/t)} \approx e^{-x-x^2/2t}, \tag{2.53}$$

hence the correction is negligible when $x \ll \sqrt{t}$, which for our problem implies $N^2 \gg \sqrt{t}$.

From Eq. (2.52), we see that the relaxation time is $\frac{N^2}{2\pi^2}$. This is plausible since in order to homogenize the distribution on the scale of the ring, we need to diffuse over N sites, which demands $O(N^2)$ steps – according to our previous results on 1D diffusion. Note that for times of the order of the relaxation time, the condition $N^2 \gg \sqrt{t}$ reduces to $N \gg 1$, thus validating the approximation previously used.

Another nice way to see the relation between the "discrete" approach and the continuous one is by considering the *difference* equation for the discrete probabilities at the sites, p_i. It is given by

$$\delta \vec{p} = \tilde{P} \vec{p}, \tag{2.54}$$

with $\tilde{P} = P - I$, I being the identity matrix.

In the continuum limit, the LHS will become a time derivative ($\delta p_i \approx \frac{\partial p}{\partial t} \tau$, with τ the duration of each step), while the RHS is proportional to the discrete version of the second derivative operator: Its ith component is given by

$$\frac{1}{2}(p_{i-1} + p_{i+1} - 2p_i) = \frac{1}{2}[(p_{i+1} - p_i) - (p_i - p_{i-1})] \approx \frac{1}{2}(p'|_{i+1} - p'|_i) \approx p''/2. \tag{2.55}$$

Hence in the continuum limit this 1D random walk leads to the diffusion equation, with $D = \frac{1}{2\tau}$.

2.6   Summary

In this chapter we introduced the important concept of a random walk – on which the next chapter will build. We used "brute-force" combinatorics to show why in the 1D case asymptotically the probability distribution of the random walker is Gaussian, with a variance that increases linearly in time – the hallmark of diffusion. We then followed Einstein's derivation to derive the diffusion equation in *any* dimension (under certain simplifying assumptions which we will relax in Chapter 7). Solving the diffusion equation showed that the insights gained from the 1D analysis remain intact. The problems at the end of this chapter further substantiate the various interesting properties and diverse applications of random walks in various spatial dimensions.

We next considered random walks on *networks*, which led us to a more general discussion of Markov chains and of the Perron–Frobenius theorem. This theorem forms the basis of the influential Google PageRank algorithm, which we also reviewed. Finally, we showed how one can relate the Markov chain formalism to the 1D diffusive behavior studied early on in this chapter, with an interesting mathematical relation to the FFT algorithm.

For further reading There are many excellent books dealing with random walks. Some useful resources are Feller (2008); Gardiner (2009); Schuss (2009); Paul and Baschnagel (2013); Hughes (1996); Krapivsky, Redner, and Ben-Naim (2010), all of which have extended discussions of many of the topics covered in this chapter.

2.7   Exercises

2.1   1D Random Walks and First Return Times*

Consider a 1D symmetric random walk, with step size a and time step T.

(a) Show that in the continuum limit (i.e., for times $\gg T$ and spatial resolution $\gg a$) the walker is described by the diffusion equation $\frac{\partial p}{\partial t} = D\frac{\partial^2 p}{\partial x^2}$, and find D.

(b) What is the probability that the walker is found precisely at the origin after the first N steps? In the continuum version, what is the probability *density* to be at the origin after time t?

(c) What is the probability that the walker is always to the right of the origin within the first N steps? *Hint*: Enumerate such paths by mapping this problem to a circular track on which are written N numbers, a fraction of which are $+1$ and the rest are -1. What is the probability that the sum of numbers along the track (say, in the clockwise direction) starting from one of the N points is always positive? Which of the N points cannot be possible starting points? [See also Kostinski, S. and Amir, A. An elementary derivation of first and last return times of 1D random walks. *American Journal of Physics* 84 (1), 57–60 (2016) for an extended discussion].

(d) Use the previous results to prove that a 1D walk is recurrent, i.e., that in the course of time the walker will return to the origin with probability 1.

(e) In the continuum limit characterized by the diffusion constant D, what is the probability distribution of the time for the walker to first return to the origin, $F(t)$? What is the average time to return to the origin?

2.2   1D Random Walks and *Last* Return Times

Consider a discrete 1D random walk, with step size a, time step T, and number of steps N.

(a) Suppose that the walk ends at a time t_{tot}. What is the probability $p(t)$ to return to the origin for the last time at some time $t < t_{tot}$? *Hint:* consider the discrete case and the results of Problem 2.1, then go to the continuum limit.

(c) Assuming that the amount by which one sports team leads another is well described by a random walk, what does the result of (a) suggest about lead changes in sports games?

[See also: Clauset, A., Kogan, M., and Redner, S., Safe leads and lead changes in competitive team sports, *Physical Review E*, 91(6), p.062815 (2015), as well as the reference pointed out in Problem 2.1(c)].

2.3   First Passage Times in 1D

Consider a particle undergoing diffusion on a 1D lattice, starting at the origin. The objective of this problem would be to find the probability distribution of the time it takes the particle to reach a site a distance $N \gg 1$ lattice constants away to the right of the origin (this is known as the First Passage Time – FPT).

(a) Set up the appropriate PDE to describe $P(x,t)$, and define the relevant boundary conditions for finding the FPT distribution.

(b) Find an explicit solution of the PDE. *Hint:* consider superimposing the solution of the PDE on the infinite 1D line with $\delta(r)$ initial condition at $t = 0$ with the solution on the infinite line with an initial condition of $A\delta(r - r_0)$, and find the appropriate values for A and r_0.

(c) Using the results of part (b), what is the probability to hit the target at any time? What is the mean time to reach the target?

2.4   Green's Function Warmup

Consider a particle with uncertain initial position following the probability distribution $p_0(x)$. Show that the position distribution at time $t > 0$ is given by a convolution of the form

$$p(x,t) = \int_{-\infty}^{\infty} \varphi(x - y,t)p_0(y)dy.$$

2.5 Biased Random Walks

(a) Repeat Einstein's logic in deriving the diffusion equation in two dimensions, *without* assuming isotropy.

(b) Solve the resulting equation.

2.6 Recurrent and Nonrecurrent Random Walks

Consider a random walker on a d-dimensional lattice.

(a) Define u as the probability to return to the origin. What is the probability to visit the origin precisely k times?

(b) What is the expected number of times to visit the origin? Show that it is finite if and only if $u < 1$.

(c) In the continuum limit, what is the probability density to be at the origin after time t?

(d) Use the results of (b) and (c) to show that 2D walks are recurrent while for $d > 2$ they are not. This is known as Pólya's theorem, named after the mathematician George Pólya.

It has been suggested that this is why when moles look for their mates they stick to a given depth underground!

2.7 Recurrent and Nonrecurrent Random Walks Revisited

(a) Consider a possibly asymmetric random walk in one dimension. A random walker starts at $x = 0$, and at every time step, the walker takes a step to the right with probability p and a step to the left with probability $q = 1 - p$, as illustrated in Fig. 2.5.

   Determine whether the random walk is recurrent. Recall that a random walk is said to be recurrent if it eventually returns to the origin with probability 1.
   Hint: Prove first that a random walk is recurrent if and only if the sum

$$\sum_{n=1}^{\infty} p_0^n$$

   is finite. Here p_0^n is the probability of the walker being at $x = 0$ at time n.

(b) Determine whether a symmetric random walk in d dimensions is recurrent.
   Hint: Since the exact expression for p_0^n is quite complicated for $d > 1$, you may wish to use an approximation.

Problem credit: Pétur Rafn Bryde.

Figure 2.5 An asymmetric random walk.

2.8 Gambler's Ruin*

Consider a discrete 1D random walk, with step size a and time step T. The walker starts at a point c with $0 < c < L$ ($L \gg a$ and $c \gg a$). In this problem we would like to find the probabilities for the walker to get to the origin or point L. Throughout the problem, it will be helpful to work in the continuum limit (see Problem 2.1(a)).

(a) Impose *absorbing* boundary conditions at 0 and at L. Once the walker reaches 0 or L, the walker cannot come back to the segment $[0, L]$. Find the eigenfunctions of the operator $\frac{\partial^2}{\partial x^2}$ on the line $[0, L]$.

(b) Write an expression for the initial probability distribution (a δ-function) as a sum of the eigenfunctions you got in part (b).

(c) Write out the probability distribution after time t ($p(x,t)$). (*Hint:* try using an ansatz where the space and time variables are separated $p(x,t) = f(x)g(t)$).

(d) Find expressions for the probability flux to exit the segment $[0, L]$ at 0 or L at time t (you may leave this as an infinite series).

(e) Find expressions for the overall probability to exit the segment at 0 or L (you may leave this as an infinite series).

(g) **Gambler's Ruin**: A gambler starts off with $100. At every round the gambler flips a fair coin, and either wins or loses $1 with equal probability. The game ends when they either have $1000 or go bankrupt. What is, approximately, the probability for each of these two events to occur? (You may either sum up the series or approximate it numerically by keeping the first few terms).

(h) Sum up the infinite series. You will find a very simple expression for the exit probability.

2.9 First Passage Time – Again!

Consider a 1D symmetric random walk, with step size a, time step T, starting at the origin.

Based on the results from Problem 2.8 (Gambler's Ruin), find the probability distribution of the time it takes the random walker to reach a site a distance X away from the origin (the first passage time). What is the mean first passage time? *Hint:* Obtain a closed-form expression for the FPT distribution by replacing the series with an integral in the appropriate limit.

2.10 Modeling Polymers as Random Walks

In this problem, we will simulate a variant of a random walk on a 2D lattice that doesn't intersect itself. At each step, the random walker can go to one of the *empty* neighboring sites; note that it can get stuck after some time, in which case we should discard the simulation. Run the simulation for a total step number of 100. Repeat as many times as you need to achieve 1000 random walks (that did not terminate). Plot the average $\langle R^2 \rangle$ and N in a log-log scale. What is the empirical power-law?

This problem is related to the self-avoiding random walks introduced by P. Flory in 1953 to model polymers. For an extended discussion, see Hughes, B. D. *Random Walks and Random Environments*, Clarendon Press, Oxford (1996).

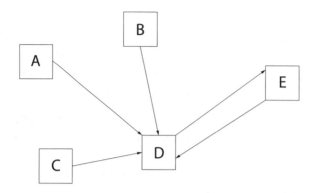

Figure 2.6 A simple network.

2.11 Scaling Approach to Solving the Diffusion Equation

In this problem we will use a different approach to solve the diffusion equation in one dimension, using scaling.

(a) Motivated by the result $\langle x^2 \rangle \propto t$, it might seem plausible to search for a solution of the form $p(x,t) = C(t)f(xt^\alpha)$. What must the value of α be? Why must the time-dependent factor $c(t)$ be included, and what must be its form?

(b) Obtain an ODE (Ordinary Differential Equation) for the function $f(z)$ by plugging the ansatz of part (a) into the 1D diffusion equation.

(c) Solve the ODE. *Hint:* it might be useful to use the simple relation $[gz]' = g'z + g$.

2.12 PageRank

Consider the web shown in Fig. 2.6.

(a) How would Google's PageRank algorithm rank pages A–E for a "damping factor" of 0.85?

(b) If a walker starts at C at $t = 0$, what is the probability of being at site C after 2 moves? 5 moves? 20 moves? How is your answer related to part (a)?

(c) Using software of your choice, simulate a "random web surfer" on pages A–F, starting at page C, and compare the frequency of visits to page C to your answers in part (b).

2.13 Hacking PageRank

Suppose we have a database with N pages that are completely linked to each other (the corresponding graph is *complete*). A spammer adds their page, probability.com, to the database and wants probability.com to be ranked as high as possible in PageRank. All pages are also linked to themselves.

(a) Add k fake pages to the database and link all of them to probability.com. There are no links between the spammer's pages and the N pages. What is the PageRank of probability.com?

(b) Hack into k of the N pages in the database and link them to probability.com. What is the PageRank of probability.com?

(c) Which of (a) and (b) gives the higher PageRank for probability.com? As an example, take $N = 100, k = 10$, and $d = 0.85$. Then, consider the case $N \gg k$.

2.14 PageRank Analysis for Matrices That Cannot Be Diagonalized*

(a) Give an example of a stochastic matrix (i.e., for which the rows sum up to one), for which the eigenvectors do not span the vector space.

(b) Consider now the PageRank scenario. By utilizing the Jordan form or Schur decomposition, prove that starting from any initial condition the probabilities to be at each site would converge to those associated with the unique $\lambda = 1$ eigenvalue (without assuming that the matrix is diagonalizable).

2.15 Markov Chains in a Board Game

Consider the board game shown in Fig. 2.7. The game board has 3×3 squares arranged in order from $i = 1$ to $i = 9$. Each player rolls a three-sided die in turn with 3 equally likely outcomes from 1 to 3 denoted by a, which is the number of steps they can move. Players do not move if $i + a > 9$. Starting from 0 off the game board, the first player to hit the last square is the winner. There are two types of magic channels on the board, which are shown as dashed and solid lines. If the player stops at the foot of the solid-line channel, they will be transported to the top of that channel. Conversely, if the player stops at the top of the dashed-line channel, they will slide down to the bottom of that channel.

(a) Having a set of states $S = \{s_i\}$, $i = 0, 1, \ldots, 9$, set up the corresponding Markov chains for the game *with* and *without* magic channels. You need to build the transition matrix T whose ijth entry t_{ij} is the probability that the Markov chain starting in state s_i will be in state s_j after one step.

(b) Consider a Markov chain having r absorbing states (i.e., that once reached are never left) and t transient (non-absorbing) states from each of which an absorbing state can be reached. Assume that the first t states are transient and the last r states are absorbing. The transition matrix is then in a block form with a lower right block of $r \times r$ identity matrix:

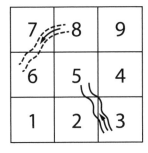

Figure 2.7 Game board.

$$T = \left(\begin{array}{c|c} Q & A \\ \hline 0 & I \end{array}\right).$$

Show that the matrix $I - Q$ is invertible and has the inverse matrix $N = I + Q + Q^2 + \cdots$. Also show that the ijth entry n_{ij} of N is the expected number of times the chain visits state s_j before getting absorbed given that it starts in s_i (where if $i = j$, the counting of visits starts from 1, i.e., n_{ii} will never be smaller than 1).

(c) Based on your results in (a) and (b), express analytically the average number of runs to complete the game *with* and *without* magic channels, and evaluate the expression numerically if needed.

Problem credit: Rui Fang.

3 Langevin and Fokker–Planck Equations and Their Applications

Probability theory is nothing but common sense reduced to calculation.

(Pierre Simon Laplace)

In our previous discussion of random walks (say in 1D), we worked with the discrete equation:

$$x_{i+1} = x_i + \xi. \tag{3.1}$$

Let us now consider the continuous version, where

$$\frac{dx}{dt} = \xi(t), \tag{3.2}$$

with $\xi(t)$ is a noise with vanishing mean and correlation function $\langle \xi(t')\xi(t) \rangle = f(t - t')$ (typically, $\langle ; \rangle$ brackets will denote averaging over all noise realizations – i.e., ensemble-averaging). This is a simple example of a *Langevin equation*, named after Paul Langevin.

In many cases we want to describe the system on timescales much larger than the correlation time of the noise – which means we can replace $f(t)$ by a δ-function, $f(t) = C\delta(t)$. Keep in mind that for certain applications, the noise correlation time could be relevant (see Problem 3.1 for an example).

Consider now the time derivative of $\langle x^2 \rangle$. For a particular realization we may express

$$x(t) = x_0 + \int_0^t \xi(t')dt'. \tag{3.3}$$

Therefore, we find

$$\frac{d \langle x^2 \rangle}{dt} = 2 \langle x(t)\xi(t) \rangle = \left\langle 2 \left[x_0 + \int_0^t \xi(t')dt' \right] \xi(t) \right\rangle. \tag{3.4}$$

Since at every t the mean of ξ vanishes, the first term vanishes and we are left with

$$\frac{d \langle x^2 \rangle}{dt} = 2 \int_0^t \langle \xi(t')\xi(t) \rangle dt' = 2 \int_0^t f(t - t')dt'. \tag{3.5}$$

Since we are integrating over *half* of the δ function, we find that

$$\frac{d\langle x^2 \rangle}{dt} = C, \tag{3.6}$$

hence $Ct = 2Dt$, and we obtain diffusion with a constant $D = C/2$.

How can we describe the random process in the δ function limit? It turns out that this can be done in several ways, which may lead to different results occasionally – two common methods used are the Itô and Stratonovich versions of stochastic calculus. The issue is that in these descriptions there are two timescales which vanish – the discretization time and the correlation time – and the order in which we send them to zero matters. We will not go deeply into this subtle issue, but for our purposes we will work mostly within the Itô formalism.

Itô vs. Stratonovich formalisms for stochastic calculus The issue of interpretation of the stochastic equations arises when dealing with the Langevin equation. Shortly, we will derive the so-called Fokker–Planck equation (named after Adriaan Fokker and Max Planck), which describes the *deterministic* evolution of the probability distribution in time – since it is "just" a PDE, there is no ambiguity of interpretation once the Fokker–Planck equation is written. For most applications we will see, both Itô and Stratonovich lead to the *same* Fokker–Planck equation. In particular, this is true when the noise is additive. However, for cases involving *multiplicative* noise (i.e., where the noise term multiplies a term associated with the variable x rather than being added to it) the resulting PDE takes a different form. In such cases, the correct stochastic interpretation depends on the particular application at hand and the relevant order of limits. For further reading on stochastic differential equations, see Gardiner (2009) and Schuss (2009).

Within the Itô formalism, the correlation time is shorter than the discretization time, so we can think of our continuous random walker as driven by a set of independent, random kicks. It is often useful to think how one would *simulate* such a process. To do this, we can make time discrete, with small time steps Δt (defining our time resolution), and for each time step we would like to choose the noise as a Gaussian, independent random variable (we will see shortly why it should be Gaussian), with magnitude W. This recovers Eq. (3.1). After time t, we would have taken $\frac{t}{\Delta t}$ steps, hence

$$\langle x^2 \rangle = \frac{t}{\Delta t} W^2. \tag{3.7}$$

To get the correct scaling, we must therefore choose $W = C\sqrt{\Delta t}$. In fact, this simple scheme to simulate stochastic processes is known as the *Euler–Maruyama* method. The discrete Langevin equation corresponding to this scheme in this situation of a simple random walk is

$$x_{t+\Delta t} = x_t + C\sqrt{\Delta t} \tag{3.8}$$

(for an interesting pedagogical paper elaborating on the utility of this discrete approach, see Gillespie 1996). What would have happened if we had chosen a non-Gaussian distribution? In that case, if we decrease our time resolution, i.e., make the step size larger, the effective noise in each step would be the sum of a large number of variables, and would therefore converge to a Gaussian (assuming that the distribution has finite variance – this assumption is relaxed in Problem 6.10!). On the other hand, Gaussian distributions are infinitely divisible – we can always write $\xi = \xi_1 + \xi_2$, where ξ_1 and ξ_2 are *also* Gaussian variables, with half the variance. So we can think of our Gaussian variable as resulting from a large sum of independent Gaussian variables.

Divisibility: can a random variable be expressed as a sum of two independent random variables? Problem 3.10 deals with the *lack* of divisibility of the uniform distribution – a uniformly chosen random variable cannot be represented as the sum of two other (independent) random variables! In Chapter 6 we will find the most general family of distributions which have the much stronger property which the Gaussian distribution exhibits – whereby the sum of two (or more) variables drawn from the distribution results in a variable corresponding to a probability distribution identical in shape to the original (i.e., the two are equal upon appropriate shifting and scaling).

Eq. (3.2) is the simplest instance of a Langevin equation, first introduced by Langevin to model diffusion of particles in a liquid or gas. We have seen before that the probability distribution is described by the diffusion equation, which is the simplest instance of a Fokker–Planck equation. Next, we will study more general Langevin equations, of the form

$$\frac{dx}{dt} = -\mu V'(x) + \xi. \tag{3.9}$$

The function $f(x) = -V'(x)$ can be associated with a physical force being applied to the diffusing particles, but in other scenarios it will be associated with a feedback in the system (we will study a biological example of this sort). Moreover, many times the Langevin equation will be defined in a higher dimension, as we shall study when we deal with certain "escape-over-a-barrier" problems.

Example I: Particle in a Potential Driven by Random Noise

Consider a particle in a potential $V(x)$ driven by random kicks due to its interaction with a thermal bath. For simplicity, let us consider the 1D case. The Langevin equation reads

$$m\frac{dv}{dt} = -\gamma v + \zeta(t) - V'(x), \tag{3.10}$$

where ζ is the noise and γ the dissipation. In fact, the two originate from the same physical mechanism (interactions with the environment) and so are physically related – as we shall shortly see.

The Ornstein–Uhlenbeck process In the case $V = 0$, Eq. 3.10 is known as the Ornstein–Uhlenbeck process. Its properties are explored within Problem 3.3.

One commonly relevant limit is the *overdamped* case, in which the particle's inertia is irrelevant (see also Appendix D). In this case we can ignore the LHS, and write

$$\frac{dx}{dt} = \mu f(x) + \xi, \tag{3.11}$$

where $f(x) = -V'(x)$, $\xi = \zeta/\gamma$ and $\mu = 1/\gamma$. This is precisely Eq. (3.9). In the case where inertia is relevant, it is convenient sometimes to think of the dynamics in a two-dimensional space, by defining a vector $\vec{x}$ with $x_1 \equiv x$, $x_2 \equiv mv$. Therefore,

$$\frac{d\vec{x}}{dt} = -\mu \nabla \tilde{V}(\vec{x}) + \vec{\psi}, \tag{3.12}$$

where in this case μ is a *matrix*

$$\mu = \begin{pmatrix} 0 & -1 \\ 1 & \gamma \end{pmatrix}, \tag{3.13}$$

$\tilde{V} = \frac{1}{2}mv^2 + V(x)$, and the noise term is given by $\vec{\psi} = (0, \gamma \xi)$. This is a generalization of Eq. (3.9) which we shall use later in the book.

We shall now understand how to go from this Langevin equation to the so-called Fokker–Planck equation, which generalizes our derivation of the diffusion equation. We shall then use the Langevin equation to study a contemporary problem in biology, and will see how analogous equations to the Langevin and Fokker–Planck equations arise in the context of economics. In Chapter 4 we will study a more involved physical problem using Fokker–Planck equations known as "escape-over-a-barrier."

Example II: Density Gradients in an Isothermal Fluid: Thermal Equilibrium Arising from Superimposing Diffusion and Drift

This may at first seem off-topic, but will end up helping us develop intuition for deriving the Fokker–Planck equation. Consider small particles immersed in a liquid, where m is the difference between the particle mass and that of the liquid it displaces (i.e., the buoyant mass). Let us also assume that the system is at thermal equilibrium at temperature T. This is precisely the setup of the classical experiment set up by Jean Baptiste Perrin at the beginning of the twentieth century (motivated by Einstein's explanations of diffusion), where he tracked individual particles diffusing in the medium using microscopy – see Fig. 2.1 for an example of a particle's trajectory.

According to the Boltzmann distribution, the density of particles at a given height h will be

$$n(h) \propto e^{-\frac{mgh}{kT}}, \tag{3.14}$$

with k the Boltzmann constant (see Appendix D for a reminder regarding the Boltzmann distribution).

Note that the above equation is for the spatial dependence of *particle density*. We could have written the same equation for the *probability distribution* of particle position – the two quantities only differ by a normalization constant.

We will now understand how this comes about from the "molecular" picture of Eq. (3.9). The particles are subject to the force of gravity, which, if the density were to be constant, would lead to a constant current of particles downward – in the overdamped limit the particles would move at constant velocity $v = \mu F = \mu mg$. However, this initial drift would lead to a higher concentration of molecules at lower altitude. The gradient of density implies that the *random walk* of the molecules would lead to an effective current *upward*. This current is proportional to the gradient of density (and opposite in sign), a result known as Fick's law. To illustrate this, let us temporarily assume that no external forces are acting on the molecules (i.e., we switched gravity "off" for now). Consider the relation

$$\vec{j} = -D\nabla n. \tag{3.15}$$

From the continuity equation we have

$$\nabla \cdot \vec{j} + \frac{\partial n}{\partial t} = 0, \tag{3.16}$$

where $\nabla\cdot$ is the divergence of the current $\vec{j}$ (which is – in the general case – a vector field). Hence

$$\frac{\partial n}{\partial t} = D\nabla^2 n, \tag{3.17}$$

and we see that the diffusion equation we previously derived indeed arises from Fick's law.

Since at steady-state all time derivatives vanish, the divergence of the current must vanish. In this effectively one-dimensional problem this implies the current must be constant. Since there can be no flux at the surface, this implies the current must vanish everywhere. In other words, the current due to diffusion (upward) would cancel that of gravity:

$$-Dn' = mgn\mu. \tag{3.18}$$

The solution is indeed that of Eq. (3.14), provided that: $D = kT\mu$. This relation is known as the Einstein relation and is extremely useful. It implies that both diffusion and μ (which is the reciprocal of the drag coefficient γ) are related – they both arise from the same physical mechanism: collisions with other molecules. This is an example of a fluctuation–dissipation theorem: It relates the diffusion constant (associated with the fluctuations) to the mobility (associated with the dissipation due to the viscous drag).

Finally, it is worth noting that from *measuring* the density profile of Eq. (3.14), the Boltzmann constant (and thus Avogadro's number) can be inferred. This was part of

Perrin's work for which he was awarded the Nobel prize in 1926 – essentially verifying Einstein's ideas and confirming the atomistic nature of matter!

Going from a Langevin to a Fokker–Planck equation Consider a particle performing a 1D random walk in the overdamped limit – as described by the Langevin equation of Eq. (3.9) (with μ potentially depending on position). The continuity equation (3.16) is generally true, and the current will have, as in the previous example, two contributions: one from the force ($\vec{j}_f$), and the second from the diffusion term ($\vec{j}_d$). The former can be readily understood in the overdamped limit in the absence of stochasticity: The force would lead the particle to a constant velocity $\vec{v} = \mu \vec{f}$, hence the current will be

$$\vec{j}_f = -\mu p \nabla V(\vec{r}). \tag{3.19}$$

In the absence of a force, we saw that the diffusion equation arises (following Einstein's derivation). Furthermore, the example in the previous section showed that it may also be interpreted as the continuity equation combined with Fick's law; therefore, the diffusive current is

$$\vec{j}_d = -D\nabla p. \tag{3.20}$$

Plugging Eqs. (3.19) and (3.20) into the continuity equation $\frac{\partial p}{\partial t} + \nabla \cdot \vec{j} = 0$, we obtain the Fokker–Planck equation,

$$\frac{\partial p}{\partial t} = \nabla \cdot [\mu \nabla V(\vec{r})p] + \nabla[D\nabla p], \tag{3.21}$$

where in general D and μ can be space-dependent, and if we are dealing with a statistical mechanics problem they are related by the Einstein relation. Note that as in Eq. (3.16), $\nabla \cdot$ denotes the divergence of the vector field. This is the general form of the Fokker–Planck equation – and is basically derived by considering the "probability currents," just as we have done when deriving the diffusion equation (the derivation can be done more systematically, within the Itô or Stratonovich stochastic calculus formalism – in this case both lead to the same equation).

Revisiting Einstein's Derivation in the Presence of a Potential

Previously, we derived the diffusion equation following Einstein's insights, while assuming that the particle is diffusing in a uniform potential. We shall now repeat the derivation for a particle in a one-dimensional potential $V(x)$, assuming the overdamped limit (where velocity is always proportional to force). Our starting point will be the discrete Langevin equation (see Eq. (3.8)),

$$x_{t+\Delta t} = \underbrace{x_t + \mu f(x_t)\Delta t}_{\text{Deterministic}} + \underbrace{\xi\sqrt{\Delta t}}_{\text{Stochastic}}, \tag{3.22}$$

with time step Δt and ξ a Gaussian with vanishing mean and standard deviation σ.

Let us assume that the probability distribution at time t is $P(x,t)$. If the particle was at position $\tilde{x}$ at time t, clearly the probability that it will be at position x at time $t + \Delta t$ is given by the probability that the noise term equals $x - \tilde{x} - \mu f(\tilde{x})\Delta t$; hence it equals

$$G(x|\tilde{x}) = \frac{1}{\sqrt{2\pi\sigma^2\Delta t}}e^{-\frac{[x-\tilde{x}-\mu f(\tilde{x})\Delta t]^2}{2\sigma^2\Delta t}}. \tag{3.23}$$

We can also approximate $f(\tilde{x}) \approx f(x) + f'(x)(\tilde{x} - x)$ (for small Δt, this first-order Taylor expansion will be a good approximation). This allows us to write an equation for the time evolution of $p(x,t)$, by considering all the possibilities of $\tilde{x}$:

$$p(x,t + \Delta t) = \int G(x|\tilde{x})p(\tilde{x},t)d\tilde{x}. \tag{3.24}$$

Following a similar logic to our previous derivation of the diffusion equation, we may now expand $p(\tilde{x},t)$ to second order (around the point x):

$$p(\tilde{x},t) \approx p(x,t) + p'(x,t)(\tilde{x} - x) + \frac{p''(x,t)}{2}(\tilde{x} - x)^2. \tag{3.25}$$

Plugging this form into Eq. (3.24), we may readily perform the Gaussian integrals and find (to order $O(\Delta t)$)

$$p(x,t + \Delta t) = \frac{p(x,t)}{1 + \mu\Delta t f'} - \mu\Delta t f p'(x,t) + \frac{\sigma^2}{2}p''(x,t). \tag{3.26}$$

Expanding to first order in Δt we have $\frac{1}{1+\mu\Delta t f'} \approx 1 - \mu\Delta t f'$, and taking the continuous limit leads to the one-dimensional version of Eq. (3.21).

Note that in Eq. (3.22) the force is evaluated at x_t. Identical results are obtained when the force is evaluated at $x_{t+\Delta t}$.

Revisiting Einstein's derivation in higher dimensions Try to repeat the derivation for a particle in a potential $V(\vec{r})$, assuming the overdamped limit. You should recover Eq. (3.21).

3.1   Application of a Discrete Langevin Equation to a Biological Problem

[Note: this section can be skipped without affecting the flow of the rest of the text]

This section deals with the application of a discrete version of a Langevin equation to a problem in contemporary cell biology. It is a problem where the biological data could only be interpreted within the framework of a mathematical model, and in which the quantitative details actually matter and lead to conceptual advances (for two additional – and very different – problems where Fokker–Planck equations are used to draw conclusions in a biologically relevant setting, see Problems 3.11 and 3.12).

A little biological background: bacteria, such as the *E. coli* strains that we all have in our guts, grow such that they double their volume in several tens of minutes (in good nutrient conditions the doubling time can be as short as 20 minutes), after which they divide symmetrically. It turns out that during this period the cell volume grows *exponentially*. This brings about an interesting problem: If we naively assume that the cell cycle relies on a "timer" (i.e., the duration between two division events is set by a clock), then denoting the cell size at birth by v_b, its size at division will be given by

$$v_d = v_b 2^{t/t_d}, \tag{3.27}$$

hence the newborn size of the daughter cell is

$$\log_2 v_b^{n+1} = \log_2 v_b^n + t/t_d - 1, \tag{3.28}$$

where the superscripts n, $n+1$ denote the generation number (n corresponding to the mother cell, and $n+1$ for the daughter cell). If $t = t_d$ precisely, we have perfect doubling and perfect division every generation, and all would be fine. However, nothing is perfect, and there must be fluctuations in t around t_d. Defining $\xi = t/t_d - 1$, Eq. (3.28) shows that the *logarithm* of size would perform a random walk. Hence size will be *unbounded*, with some cells becoming very small while others very large – which is not what happens in reality.

This goes to show that there must be a feedback mechanism that (potentially among other functions) controls size. The goal of this section would be to derive a phenomenological description of this feedback, in the spirit of the Langevin equation. Such a model would go beyond the "timer" described above, converging to a stable size distribution in the same way that the restoring force in the Ornstein–Uhlenbeck process of Eq. 3.10 leads to a stationary velocity distribution (and without it, as we have seen, the velocity will perform a random walk – analogous to the random walks of the logarithm of cell size discussed above). Such a model can make predictions regarding the statistics of the model variables (namely, cell size and time), including correlations between cells in the same lineage and distributions of these variables. These relations can then be empirically tested, to validate or falsify the model.

Formally, we can describe this using a discrete version of the Langevin equation. The notation for the discussion that follows is summarized in Table 3.1.

3.1.1   Deriving a Discrete Version of the Langevin Equation to Describe Size Control

We will assume that the cell controls size by attempting to divide at a size

$$v_d = f(v_b). \tag{3.29}$$

The choice of function f determines the regulation strategy used by the cell. For example, the (failed) timer strategy we described above corresponds to $f(v_b) = 2v_b$. An a priori reasonable strategy is $f(v_b) = const$, in which the cell attempts to divide when reaching a critical size (this thresholding strategy is known as a "sizer"). Clearly,

Table 3.1 Notations for cell size regulation problem

Variable	Definition
v_b	Size at birth (stochastic variable)
v_d	Size at division (stochastic variable)
v_0	Average size at birth
n	Subscript or superscript denoting generation number
$f(x)$	Size regulation (deterministic) strategy
t_d	Volume doubling time (assumed constant)
ξ	Time-additive noise
t_a	Deterministic generation time given a size at birth
t_n	Noise added to deterministic generation time
σ_n	Standard deviation of t_n
x	$\log_2[v_b/v_0]$, see Eq. (3.34)
σ_v	Standard deviation of x for the stationary solution
σ_t	Standard deviation of generation time t for the stationary solution
C_{xy}	Pearson correlation coefficient between variables x and y

this will have strong size control (i.e., size distributions will be as narrow as possible), but we shall later claim it is also not consistent with experimental data on *E. coli*. In order to implement a strategy described by a function f, the cell must attempt to grow for a time

$$t_a = t_d \log_2(f(v_b)/v_b). \tag{3.30}$$

To this "deterministic" component we will add a stochastic term t_n such that the actual generation time t will be given by

$$t = t_a + t_n. \tag{3.31}$$

This is precisely the philosophy which Langevin used in writing his equation, where the force due to collisions with the molecules was decoupled into a deterministic component (leading to the viscous drag) and a noise term. We will assume the noise t_n to be Gaussian with vanishing mean and standard deviation σ_n (though this assumption can be relaxed without affecting the main conclusions). This concludes the definition of the model (for now). To make further progress, we will assume that the cell size distribution will ultimately be narrow (an assumption which can be verified experimentally, and also self-consistently within our model). Hence we don't actually care about the behavior of $f(x)$ for any x, but really only for a narrow range around the typical newborn cell size v_0. This implies that we can Taylor expand f around v_0 to find that:

$$f(v_b) \approx 2v_0 + f'|_{v_0}(v_b - v_0), \tag{3.32}$$

where we utilized the fact that for the typical newborn size v_0 we should have $f(v_0) = 2v_0$, such that cells double in volume.

It will be useful to denote $f'|_{v_0} = 2(1 - \alpha)$. In this notation the "timer" strategy corresponds to a slope of 2 and hence $\alpha = 0$, while the thresholding "sizer" strategy corresponds to vanishing slope and hence $\alpha = 1$. For any f, we can find the relevant value of α by evaluating its derivative – for example, if $f(v_b) = v_b + v_0$, then $\alpha = \frac{1}{2}$.

All models with the same value of α will follow approximately the same behavior, since their Taylor expansion agrees to first order. To solve the model, it will be useful to choose a particular function $f(v_b) = 2v_b^{1-\alpha}v_0^{\alpha}$. You can check that its derivative at v_0 gives $2(1-\alpha)$, hence the α appearing in this definition of f is indeed consistent with our previous definition of α. This function of f was chosen such that the "attempted" generation time will take a particularly simple expression:

$$t_a = t_d - t_d \alpha \log_2[v_b/v_0]. \tag{3.33}$$

Defining $x \equiv \log_2[v_b/v_0]$, our stochastic equation becomes

$$x_{n+1} = x_n(1 - \alpha) + \xi, \tag{3.34}$$

with ξ a Gaussian noise with standard deviation σ_n/t_d. This is a discrete version of the Langevin equation we studied earlier in this chapter! In fact, it looks a lot like the Ornstein–Uhleneck process mentioned at the beginning of this chapter and explored in Problem 3.3 (see Ho, Lin, and Amir (2018) for an extended discussion). If $\alpha = 0$, we are back to the random walk scenario. $\alpha > 0$ corresponds to a feedback, which *might* be able to correct the random drift due to the noise term ξ.

3.1.2 Finding the Stationary Size and Time Distributions from the Discrete Langevin Equation

It is very easy to simulate this equation – which Problem 3.6 deals with. But it is also easy to guess a solution – since ξ is a Gaussian variable with vanishing mean, if x_n is Gaussian with vanishing mean x_{n+1} will also be such, so we would have a stationary solution if the variances will also be consistent. Denoting the variance of x_n by σ_v^2, we obtain the self-consistent equation

$$\sigma_v^2 = (1 - \alpha)^2\sigma_v^2 + (\sigma_n/t_d)^2. \tag{3.35}$$

Note that we relied on the fact that the noise term is uncorrelated with the size at the beginning of that generation (x_n), but it *will* be correlated with the size at the end of that generation (x_{n+1}), and the two variables x_n, x_{n+1} may also be correlated.

Solving Eq. 3.35, we find that

$$\sigma_v^2 = (\sigma_n/t_d)^2/[\alpha(2 - \alpha)]. \tag{3.36}$$

Hence we see that as $\alpha \to 0$, the size distribution becomes infinitely broad, as we expect from our consideration of the case $\alpha = 0$. Less obvious a priori is that as $\alpha \to 2$ the same happens. Therefore, the regime of stable size distributions is $0 < \alpha < 2$, and it is easy to see that the distribution is narrowest for $\alpha = 1$, which is also intuitive, since that corresponds to the size-thresholding case.

Furthermore, since the logarithm of size was found to be a Gaussian variable, the size distribution will be *log-normal*:

$$p(v_b) \propto e^{-\log_2(v_b/v_0)^2/\sigma_v^2}/v_b. \tag{3.37}$$

Similarly, we can find the time distribution. Since the time is given by

$$t = t_a + t_d \xi, \tag{3.38}$$

and t_a was found to be Gaussian, the time distribution will also be Gaussian, with a variance of

$$\sigma_t^2 = \alpha^2 t_d^2 \sigma_v^2 + \sigma_n^2 = \sigma_n^2(1 + \alpha^2/[\alpha(2-\alpha)]) = 2\sigma_n^2/(2-\alpha). \tag{3.39}$$

As we might expect, the time distribution is *narrowest* for $\alpha = 0$ (the "timer" is good at controlling time!), and monotonically increases with α. It is important to note that one stochastic noise term, σ_n, determines the width of both *size and time distributions*. It is useful to quantify the relative width of distributions by their *coefficient of variation* (CV), which is the ratio of their standard deviation to their mean. Using Eqs. (3.37) and (3.39) it is easy to see that the CV of the size distribution is $\log(2)/\sqrt{2\alpha}$ times that of the time distribution.

3.1.3   Quantifying Correlations between Variables and Collapsing the Size and Time Distributions

Armed with this machinery, we can now quantify correlations between various variables – that can readily be compared with experimental data. A useful entity is the Pearson correlation coefficient (named after Karl Pearson), which is defined as the following expectation value

$$C_{xy} \equiv \frac{\mathbf{E}[(x - \bar{x})(y - \bar{y})]}{\sigma_x \sigma_y}, \tag{3.40}$$

where $\bar{x}, \bar{y}$ are the averages of the variables, and σ_x, σ_y their standard deviations.

Clearly, the Pearson correlation coefficient of a variable with itself is 1, and with its negative is -1. Additionally, the correlation coefficient vanishes for two independent variables. Therefore, its value between -1 and 1 is a measure of how correlated or anti-correlated two variables are. Furthermore, it is clear that this definition is insensitive to adding a constant to the variables, or scaling by a positive factor.

Consider now the size of a cell at birth, and its size at division, $C_{v_b v_d}$. According to our assumption of perfectly symmetric division, the correlation coefficient between these two variables would be the same as that between size at birth and the size of the daughter cell at birth. Plugging into the definition we have

$$C_{v_b v_d} \equiv \frac{\mathbf{E}[(v_b^n - v_0)(v_b^{n+1} - v_0)]}{\sigma^2}, \tag{3.41}$$

with σ^2 the variance of the birth size. To simplify things, it is useful to note that by Taylor expanding $x = \log_2(v_b/v_0)$ around v_0, the correlation coefficient would be nearly the same for the variables x_n and x_{n+1}, hence

$$C_{v_b v_d} \approx \frac{\mathbf{E}[x_n x_{n+1}]}{\sigma_v^2}. \tag{3.42}$$

Using Eq. (3.34), and the fact that the noise is uncorrelated with x_n, we find that

$$C_{v_b v_d} \approx \frac{\mathbf{E}[x_n^2(1-\alpha)]}{\sigma_v^2} = 1 - \alpha. \tag{3.43}$$

For $\alpha = 1$, the vanishing of the correlation coefficient is very plausible: since in this case thresholding washes away the memory of the initial size. In the case $\alpha = 0$, which has no size control, the cell attempts to double its size, which is why we get a perfect correlation coefficient.

This correlation coefficient was experimentally measured by Koppes *et al.* (1980), and found to be approximately 1/2, suggesting that $\alpha = 1/2$. As mentioned earlier, this is what we would get if $f(v_b) = v_b + v_0$, i.e., the cell attempts to add a relative volume to its volume at birth. But note that this can only be interpreted within a quantitative model such as the one we described above, which is the reason for which this "interpretation" of the measurement was not provided until 2014, several decades after the experimental work (Amir 2014). We can now calculate additional correlation coefficients to further test the theory, for example, that between the size at birth and the time to division. As before, it is useful to replace v_b by x, which hardly affects the coefficient. Thus

$$C_{v_b t} \approx \frac{\mathbf{E}[x(t - t_d)]}{\sigma_v \sigma_t} = \frac{\mathbf{E}[x(t_a - t_d)]}{\sigma_v \sigma_t}. \tag{3.44}$$

Using $t_a - t_d = -t_d \alpha x$, we find that

$$C_{v_b t} \approx -\alpha \frac{t_d \mathbf{E}[x^2]}{\sigma_v \sigma_t} = -\alpha \frac{\sigma_v t_d}{\sigma_t}. \tag{3.45}$$

Using Eqs. (3.36) and (3.39) we find that

$$C_{v_b t} \approx -\alpha t_d \sqrt{\frac{(\sigma_n/t_d)^2/[\alpha(2-\alpha)]}{2\sigma_n^2/(2-\alpha)}}. \tag{3.46}$$

Hence

$$C_{v_b t} \approx -\sqrt{\alpha/2}, \tag{3.47}$$

and for $\alpha = 1/2$ we expect it to be about $-1/2$. Precisely this value was reported in Robert *et al.* (2014), corroborating this model. Finally, the fact that one source of stochasticity determines both the size and time distributions allows us to scale both distributions without using any fitting parameters: From Eqs. (3.36) and (3.39), we find that for $\alpha = 1/2$, the distribution of the scaled size variable $\log_2(v_b/v_0)$ and the scaled time variable $(t - \tau)/\tau$ both have vanishing mean and identical variance. This is shown in Fig. 3.1, from Soifer, Robert, and Amir (2016). As a corollary of this, CV of the size distribution is $\log(2)$ smaller than that of the time distribution, which was also corroborated experimentally in Soifer, Robert, and Amir (2016).

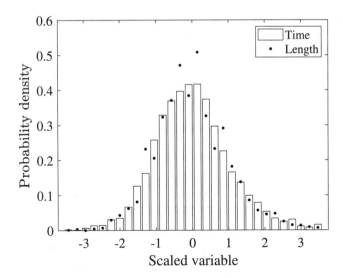

Figure 3.1 Data collapse of the scaled size distribution (dots) and generation time distribution (bar graph) for the bacterium *E. coli*, without any fitting parameters. Adapted from Soifer, Robert, and Amir (2016).

In fact, the story is more involved, and there are other models that are not "birth-centric" which would give the same correlations, but that is outside the scope of this book. Interested readers may refer to Ho and Amir (2015) for a discussion and more biological insights into this problem, and Ho, Lin, and Amir (2018) for further details of mathematical models of cell size control in microbes.

3.2   The Black–Scholes Equation: Pricing Options

[Note: this section can be skipped without affecting the flow of the rest of the text]
Next, we introduce some basic ideas of option pricing in finance, and present a derivation of the Black–Scholes equation as the continuous limit of a discrete-time model. This section is inspired by the notes of Terence Tao.[1]
We focus on two primary options, the **call** and **put options**, which are in a sense inverse to each other. We also only study **European options**, which may only be exercised at the expiration time, denoted t_1 (an American option, on the other hand, allows the owner to exercise at any time prior to the expiration time). Throughout, we consider some underlying asset S, with price S_t at time t.

1. **Call option**: Right to purchase S at price P (strike price) from the seller of the option at time t_1.
2. **Put option**: Right to sell S at price P to the seller of the option at time t_1.

[1] T. Tao, "The Black-Scholes Equation." https://terrytao.wordpress.com/2008/07/01/the-black-scholes-equation.

Table 3.2 Notations for the Black–Scholes problem

Variable	Definition
r	Interest rate
S_t	Stock price at time t
μ	Drift of stocks, see Eq. (3.50)
σ	Volatility of stock dynamics, see discussion following Eq. (3.50)
$V(t, S)$	Option price at a given time and given the current stock price (for Call or Put option)
t_1	Expiration time, at which Call/Put option are exercised
P	Strike price (for Call or Put option)

The central option pricing problem is as follows: at time $t_0 < t_1$, what is the appropriate price to assign to an option with expiration date t_1 with strike price P?

To solve this problem, we make a number of simplifying assumptions:

1. **No arbitrage**: This is the central assumption we will rely on for computing prices; essentially, it means that we cannot make "easy money" – say by buying in euros and selling in dollars, etc.
2. **Infinite liquidity**: Cash is not a constraint; participants can buy and sell options at any time and price.
3. **No market power**: The buying/selling of an asset has no effect on its price.
4. **No transaction costs**.
5. **Infinite credit**: Participants can borrow or lend as much as desired at a given interest rate r.
6. **Infinite divisibility**: The asset and option can be sold in as small an increment as desired.
7. **Short-selling**. We can sell stocks that we don't have, and later on purchase them to cover this. In a sense, it will allow us to work with *negative* stock values.

The price of an option within this model will depend on our assumptions regarding the (stochastic) time dynamics of the stocks – as will be discussed below. Table 3.2 summarizes the notation we will be using throughout the chapter.

3.2.1   Time is Money: The Time Value of Money

Our assumption of an interest rate r, together with the assumption of no arbitrage, allows us to find (for example) how much a bond that would pay us 1 dollar at a future time t_1 would be worth today, at time $t_0 = t_1 - t$: Let us call this value X. We could always borrow an amount of money X *today*, and buy such a bond, getting back 1 dollar in the future and pay off the debt then (an amount Xe^{rt}, due to interest) – since we cannot make money with this strategy (no arbitrage), it means that $1 - Xe^{rt} \leq 0$, i.e., $X \geq e^{-rt}$. On the other hand, if $X > e^{-rt}$, then I can create such a bond today and sell it, getting X units of cash, and pay a dollar to the buyers at time t_1. This way I made $Xe^{rt} - 1$ amount of cash, which is positive by assumption – hence $X = e^{-rt}$.

3.2.2 Bounding the Price of a Put Option

Let $V_t(S_t)$ denote the price of an option of asset S at time t. For example, consider a put option on S at strike price P with expiration time t_1. To avoid an arbitrage opportunity, the price of the option must reflect both the price of the asset and the time-discounted strike price. If the strike price is high and the option is cheap, we could make money from buying the option and selling the stock – in contrast to the no arbitrage assumption. Let us quantify this:

We would be buying a unit of stock and the option, at the present. The amount of money we would be making is P, but this is at a *future* time t_1, hence when comparing it to the money we pay now we have to multiply by $e^{-r(t_1-t_0)}$ (taking into account the time value of money).

Hence our *gain* is

$$X = e^{-r(t_1-t_0)}P - S_{t_0} - V_{t_0}(S_{t_0}). \tag{3.48}$$

This *cannot* be positive – otherwise we have arbitrage! Thus we conclude that

$$V_{t_0}(S_{t_0}) \geq e^{-r(t_1-t_0)}P - S_{t_0}. \tag{3.49}$$

We found a lower-bound for the option price, but can we find a precise equation for the option price? Remarkably, the answer is yes – after we make some assumptions regarding the dynamics of the stocks.

3.2.3 Stocks Make a Multiplicative Random Walk

For simplicity, let us assume that time is discrete, and call the unit of time dt. To derive the Black–Scholes equation, we will assume that

$$S_{t+dt} = S_t e^{\mu dt + \xi}, \tag{3.50}$$

where μ is a constant describing the drift of the random walk, and ξ a noise term (e.g., Gaussian), with standard deviation ξ_0 and vanishing mean. For a given dt, the logarithm of the stock will do a random walk. The size of each step has a variance ξ_0^2, and the number of steps until time t is t/dt. Therefore, the standard deviation of the logarithm of the stock will be $\xi_0\sqrt{t/dt}$. As we have seen before, in order for the limit of $dt \to 0$ to make sense, we need to scale the noise appropriately: $\xi_0 \propto \sqrt{dt}$. For this scaling, the standard deviation of $\log(S)$ will be proportional to $\sqrt{t}$. We will denote the proportionality constant by σ, which is called the **volatility** of the stock.

Note that the noise plus drift term *multiplies* the current value of the stock – hence the noise is multiplicative rather than additive. One benefit of this formulation is that the stock price will always remain positive (since we are always multiplying it by a constant close to 1). This is also known as **geometric Brownian motion** or **geometric random walk**.

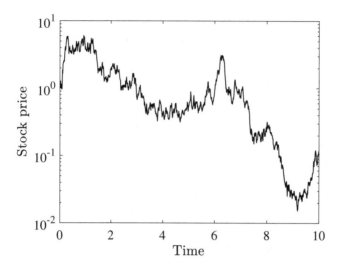

Figure 3.2 Stock price is assumed to follow a multiplicative random walk.

We can readily generate a random instance of such dynamics by running the following MATLAB code:

```
dt=0.01; mu=0.1; sigma=1; t=10;
S=1;
for indx=1:(t/dt)
      S(indx+1)=S(indx)*exp(mu*dt+randn*sigma*sqrt(dt));
end;
```

Fig. 3.2 shows the result of this code.

In fact, for the derivation below it will be easier to assume that at each discrete time step the stock price follows one of two *discrete* options, i.e., the noise will be bimodal $\pm\sigma\sqrt{dt}$. We can always "coarse-grain" this model by pooling together $N \gg 1$ steps, leading to a Gaussian noise with the same volatility, so we will not miss much by following this (simpler) route.

3.2.4   Pricing a Call Option

Suppose we are now at time t_1, and the call option (the right to purchase S at price P) is still available (i.e., suppose we are at 11:59 PM before the date of the option). If $V_{t_1}(S_{t_1}) > S_{t_1} - P > 0$, then anyone can buy S at price S_{t_1}, sell such an option at $V_{t_1}(S_{t_1})$, and sell S back at P (this hinges on the assumption $P < S_{t_1}$, otherwise we won't be able to sell S back). Thus, for the case $P < S_{t_1}$ this arbitrageur ends up with

$$V_{t_1}(S_{t_1}) - S_{t_1} + P > 0, \tag{3.51}$$

so we must have $V_{t_1}(S_{t_1}) \le S_{t_1} - P$ by the no arbitrage condition. But looking at it from the opposite point of view (i.e., short-sell S, buy an option, buy S back at P), we see that $V_{t_1}(S_{t_1}) \geq S_{t_1} - P$. Thus, in fact we must have $V_{t_1}(S_{t_1}) = S_{t_1} - P$ in the case that $S_{t_1} > P$. But if on the other hand $S_{t_1} \leq P$, then no one would buy the option at any positive price (and option prices must be nonnegative), and so the price of the option must be 0 (trivially). Thus, for a call option

$$V_{t_1}^{call}(S_{t_1}) = \max(S_{t_1} - P, 0), \tag{3.52}$$

and analogous logic shows that for a put option

$$V_{t_1}^{put}(S_{t_1}) = \max(P - S_{t_1}, 0). \tag{3.53}$$

3.2.5 Deriving the Black–Scholes Equation

Consider a call option to buy a unit of S at price P at time t_1. We will denote the price of the option by $\mathbf{V(t, S)}$. The time t can be any time ranging from the present to t_1, and S will be the price of the stock at that time. Our goal will be to show that under our assumptions we can find a unique formula for V. First, consider $\mathbf{V(t_1, S)}$. That's precisely the exercise we did in the previous section. Hence

$$\mathbf{V(t_1, S)} = max(S - P, 0). \tag{3.54}$$

We considered above the case when we could buy and exercise an option at the same time, but the more realistic and interesting case is when we purchase an option at some earlier time t_0 and have the right to exercise it at a later time t_1. We will do this recursively by considering one discrete time step back, i.e., by dt; that is, we consider buying the option at $t - dt$, *assuming* that we know how to price the option at time t. Eq. (3.54) will provide the necessary boundary condition to solve the recursive equations we will derive.

Assume that at time $t - dt$, the stock has price S_{t-dt}. According to our random walk formulation, at time t the stock can have one of two prices:

$$\frac{S_t}{S_{t-dt}} = e^{\mu dt \pm \sigma dt^{1/2}}. \tag{3.55}$$

Thus, let us define

$$S_+ = S_{t-dt}e^{\mu dt + \sigma dt^{1/2}} \approx S_{t-dt}[1 + \tilde{\mu}dt + \sigma dt^{1/2}], \tag{3.56}$$

and

$$S_- = S_{t-dt}e^{\mu dt - \sigma dt^{1/2}} \approx S_{t-dt}[1 + \tilde{\mu}dt - \sigma dt^{1/2}], \tag{3.57}$$

where we expanded the exponential retaining terms of order up to dt, and defined $\tilde{\mu} \equiv \mu + \sigma^2/2$. According to our assumptions, we know that at time t the option will be worth either $V(t, S_+)$ or $V(t, S_-)$ – but we cannot be sure which one it will be (and usually, these two will be different). Is it possible to take action at time $t - dt$ in a risk-free fashion? I.e., that our financial fate will be the same regardless of whether the stock goes up or down? For the call option under consideration, if we buy the option at $t - dt$ it is in our best interest to have the stock go up in value (since we have the option to buy it at the strike price, regardless of the stock price). How can we "hedge-away" the risk associated with the scenario in which the stock goes *down*? The solution is simple: We need to buy a negative amount of stock (this is possible, since we can short-sell). Let us find out how many negative units of stock we should buy, which we denote by x, to counteract the price change of one unit of options (we are creating a "portfolio"). If the stock goes down, we lose $V(t - dt, S)(1 + rdt) - V(t, S_-)$, and we gain $x(S(1 + rdt) - S_-)$, where r is the interest rate. Hence our total gain in the case that the stock goes down is

$$G_- = x(S(1 + rdt) - S_-) - (V(t - dt, S)(1 + rdt) - V(t, S_-)). \tag{3.58}$$

Similarly, if the stock goes up our gain will be

$$G_+ = x(S(1 + rdt) - S_+) - (V(t - dt, S)(1 + rdt) - V(t, S_+)). \tag{3.59}$$

Equating the two implies that we have *hedged-away* the risk: No matter what the stock does, our gain will be $G_+ = G_- \equiv G$! We can easily see that this amounts to choosing x obeying the equation:

$$x(S_+ - S_-) = V(t - dt, S)(1 + rdt) - V(t, S_-) - V(t - dt, S)(1 + rdt) + V(t, S_+). \tag{3.60}$$

Clearly, G cannot be positive – since then we violate the no arbitrage condition since we can deterministically make profit. On the other hand, if $G < 0$ we can buy $-x$ amount of stock, and *short-sell* an option at time $t - dt$, thereby making a positive profit of $-G$. The conclusion is that the no arbitrage condition implies that $G = 0$. Since we found x, this allows us to connect $V(t - dt, S)$ to $V(t, S_+)$ and $V(t, S_-)$! Hence we have

$$x = \frac{V(t, S_+) - V(t, S_-)}{S_+ - S_-}, \tag{3.61}$$

and

$$V(t - dt, S)(1 + rdt) = V(t, S_-) + x(S(1 + rdt) - S_-). \tag{3.62}$$

The last two equations are already sufficient to recursively find V at an earlier time – and hence we have essentially solved the problem. Here is a short piece of code which computes this, by filling in a full "column" of the matrix $V(t, s)$ based on the previous one (this is, essentially, dynamical programming). Note that for simplicity we chose $r = 0$.

```
dt=0.01; t=0:dt:1; sigma=1; mu=0;
p=1; %stock strike price
s_vec=0:0.01:50; %possible values of stock price
v=zeros(length(t),length(s_vec));
v(1,:)=max(0,s_vec-p); %boundary conditions for solving the backwards
                        recursive relation
for indx=2:length(t)
    for s_indx=1:length(s_vec)
        s=s_vec(s_indx);
        s_p=s*exp(mu*dt+sigma*sqrt(dt)); %if the market goes up
```

```
s_m=s*exp(mu*dt-sigma*sqrt(dt)); %and down
[tmp,indx_p]=min(abs(s_p-s_vec)); %used to round s_p to value
                                    in s_vec
[tmp,indx_m]=min(abs(s_m-s_vec));
x=(v(indx-1,indx_p)-v(indx-1,indx_m))/(s_p-s_m);
v(indx,s_indx)=v(indx-1,indx_p)-x*(s_p-s); %option pricing for
                                            earlier time
     end;
   end;
range=1:200;
imagesc(s_vec(range),t,v(:,range));
ylabel('Time'); xlabel('Stock price');
figure;
plot(s_vec(range),v(1,range),s_vec(range),v(end,range),'linewidth',2);
xlabel('Stock price'); ylabel('Option price'); legend('t=0', 't=1');
```

Fig. 3.3 illustrates the behavior of the option pricing according to these equations, for some arbitrary choice of parameters. Note that we are only plotting a subset of the output matrix – one needs to be careful to use only values for which the results are independent of the cutoff used for the vector of stock price and the time step used. This is why the plot in the above code restricts the range of the stock variable.

We see that as we go back in time, the sharp bilinear behavior that is encapsulated in Eq. (3.54) gets "smeared out." We will shortly understand why this is the case, based on the results of Chapter 2. Plugging Eq. (3.61) into Eq. (3.62) we find that

$$V(t-dt,S) \approx V(t,S_-) - rdt\,V(t,S_-) + \frac{V(t,S_+) - V(t,S_-)}{S_+ - S_-}[S - S_-/(1+rdt)].$$

$$(3.63)$$

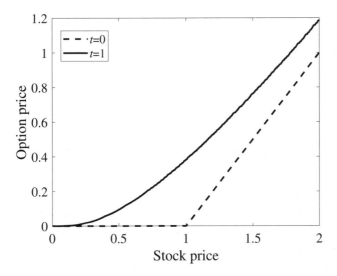

Figure 3.3 Option price (as a function of stock price) at the strike time and one unit of time beforehand.

Hence, defining $dS = S\sqrt{dt}\sigma$ and retaining terms to order $O(dt)$:

$$V(t - dt, S) \approx V(t, S) + [S\tilde{\mu}dt - dS]\frac{\partial V}{\partial S}\bigg|_{S-dS/2} - rdt\,V(t, S)$$

$$+ \frac{2dS\frac{\partial V}{\partial S}\big|_S}{2dS}[Srdt + dS - S\tilde{\mu}dt]. \tag{3.64}$$

Note that the derivative is evaluated at the mid-point between S and $S - dS$, making the approximation correct to higher order. This is essential here, since the prefactor of this term is of order $\sqrt{dt}$. Also, note that the two terms corresponding to $\pm\tilde{\mu}dt\,S\frac{\partial V}{\partial S}$ cancel each other – so that the drift term $\tilde{\mu}$ dropped out (can you see why the fact that they are evaluated at slightly different stock prices does not matter for this cancelation?).

This leads to

$$V(t - dt, S) \approx V(t, S) - dS\frac{\partial V}{\partial S}\bigg|_{S-dS/2} - rdt\,V(t, S) + \frac{\partial V}{\partial S}\bigg|_S dS + \frac{\partial V}{\partial S}\bigg|_S Srdt. \tag{3.65}$$

We can replace

$$\frac{\partial V}{\partial S}\bigg|_S dS - dS\frac{\partial V}{\partial S}\bigg|_{S-dS/2} \approx dS\frac{\partial^2 V}{\partial S^2}\bigg|_S \frac{dS}{2} = \sigma^2 dt\frac{S^2}{2}\frac{\partial^2 V}{\partial S^2}\bigg|_S, \tag{3.66}$$

leading to

$$-\frac{\partial V(t, S)}{\partial t} = -rV(t, S) + \frac{\partial V}{\partial S}\bigg|_S Sr + \frac{\sigma^2 S^2}{2}\frac{\partial^2 V}{\partial S^2}\bigg|_S. \tag{3.67}$$

This is the celebrated Black–Scholes equation, named after Fischer Black and Myron Scholes (1973). It is rather counterintuitive that the drift term μ does not appear here – only the volatility is important for pricing the option within this framework!

3.2.6   Solving the Black–Scholes Equation Using Green's Function of the Diffusion Equation

As shown above, it is straightforward to solve the equation numerically. Having cast it as a PDE, however, allows us to obtain explicit solutions. We would like to make a change of variables to cancel the interest rate r and get rid of the multiplicative nature of the random walk.

We will attempt to make a transformation of the form:

$$s = \log(S); \tilde{t} = -t; V(t, S) \to v(\tilde{t}, s). \tag{3.68}$$

Computing the partial derivatives we find that

$$\frac{\partial v}{\partial s} = \frac{\partial V}{\partial S} S, \tag{3.69}$$

and

$$\frac{\partial^2 v}{\partial s^2} = \frac{\partial^2 V}{\partial S^2} S^2 + S \frac{\partial V}{\partial S}. \tag{3.70}$$

The time derivative is

$$\frac{\partial v}{\partial \tilde{t}} = -\frac{\partial V}{\partial t}. \tag{3.71}$$

Putting it together we find

$$-\frac{\partial V(t,S)}{\partial t} = \frac{\partial v}{\partial \tilde{t}} = -rv + \frac{\partial v}{\partial s} r + \frac{\sigma^2}{2} \left[\frac{\partial^2 v}{\partial s^2} - \frac{\partial v}{\partial s} \right]. \tag{3.72}$$

This already is a simplification since all the coefficients are now constant.
The second step will be in replacing $v \to \tilde{v} = e^{r\tilde{t}} v$. This gives

$$\frac{\partial \tilde{v}}{\partial \tilde{t}} = e^{r\tilde{t}} \frac{\partial v}{\partial \tilde{t}} + r\tilde{v}, \tag{3.73}$$

hence

$$e^{r\tilde{t}} \frac{\partial v}{\partial \tilde{t}} = \frac{\partial \tilde{v}}{\partial \tilde{t}} - r\tilde{v} = -r\tilde{v} + \frac{\partial \tilde{v}}{\partial s} r + \frac{\sigma^2}{2} \left[\frac{\partial^2 \tilde{v}}{\partial s^2} - \frac{\partial \tilde{v}}{\partial s} \right]. \tag{3.74}$$

So we succeeded in eliminating the rv term:

$$\frac{\partial \tilde{v}}{\partial \tilde{t}} = [r - \sigma^2/2] \frac{\partial \tilde{v}}{\partial s} + \frac{\sigma^2}{2} \frac{\partial^2 \tilde{v}}{\partial s^2}. \tag{3.75}$$

Finally, to eliminate the first derivative term, we need to make a less intuitive substitution that will mix the time and stock variables. Defining $x \equiv s + A\tilde{t}$, with $A \equiv r - \sigma^2/2$, $\tilde{v}(\tilde{t}, s) \to \hat{v}(\tilde{t}, x)$, we find that

$$\left. \frac{\partial \tilde{v}(\tilde{t},s)}{\partial \tilde{t}} \right|_s = \left. \frac{\partial \hat{v}(\tilde{t},x)}{\partial \tilde{t}} \right|_x \left. \frac{\partial \tilde{t}}{\partial \tilde{t}} \right|_s + \left. \frac{\partial \hat{v}(\tilde{t},x)}{\partial x} \right|_{\tilde{t}} \left. \frac{\partial x}{\partial \tilde{t}} \right|_s = \frac{\partial \hat{v}(\tilde{t},x)}{\partial \tilde{t}} + A \frac{\partial \hat{v}}{\partial x}, \tag{3.76}$$

and similarly

$$\left. \frac{\partial \tilde{v}(\tilde{t},s)}{\partial s} \right|_{\tilde{t}} = \left. \frac{\partial \hat{v}(\tilde{t},x)}{\partial x} \right|_{\tilde{t}} \left. \frac{\partial x}{\partial s} \right|_{\tilde{t}} + \left. \frac{\partial \hat{v}(\tilde{t},x)}{\partial \tilde{t}} \right|_{\tilde{t}} \left. \frac{\partial \tilde{t}}{\partial s} \right|_{\tilde{t}} = \frac{\partial \hat{v}(\tilde{t},x)}{\partial x}. \tag{3.77}$$

A note on notation: in the above two equations $\Big|_a$ implies keeping the variable a fixed while taking the partial derivative with respect to the other variable – not to be confused with the notation used in our derivation of the Black–Scholes equation where this implied evaluating the partial derivative at a particular value.

Hence

$$\frac{\partial \hat{v}(\tilde{t}, x)}{\partial \tilde{t}} + A \frac{\partial \hat{v}}{\partial x} = [r - \sigma^2/2] \frac{\partial \hat{v}}{\partial x} + \frac{\sigma^2}{2} \frac{\partial^2 \hat{v}}{\partial x^2}. \tag{3.78}$$

So we end up with the standard diffusion equation:

$$\frac{\partial \hat{v}(\tilde{t}, x)}{\partial \tilde{t}} = \frac{\sigma^2}{2} \frac{\partial^2 \hat{v}}{\partial x^2}. \tag{3.79}$$

This allows us to use our knowledge and intuition from studying random walks also for this problem! Previously, we solved this equation for a δ-function boundary condition (i.e., Green's function or propagator). We can easily recast our initial conditions for $V(t_1, S)$ to one in terms of x, $\tilde{V}$, which will denote $V_0(x)$.

The solution for $\hat{v}$ can now be readily expressed in terms of the Green function:

$$G(t, t', x, x') = \frac{1}{\sqrt{4\pi D|t - t'|}} e^{-\frac{|x-x'|^2}{4D|t-t'|}}. \tag{3.80}$$

Thus

$$\hat{v}(\tilde{t}, x) = \int G(\tilde{t}, 0, x, x') V_0(x') dx'. \tag{3.81}$$

The integrals can be evaluated in terms of the error function, and the solution is known as the Black–Merton–Scholes formula (Black and Scholes 1973; Merton 1973) (Robert C. Merton and Myron Scholes received the Nobel prize for their theory of option pricing).

3.3   Another Example: The "Well Function" in Hydrology

A mathematically similar example where a diffusion equation is solved using Green's function is found in the field of hydrology (Theis 1935): Consider pumping from a deep well, connected to an underwater reservoir, which is (approximately) two-dimensional (i.e., it is thin compared to its lateral extent). It turns out that the porous nature of the soil implies that to a good approximation the velocity is proportional to the pressure gradient (this is known as "Darcy's law"). This helps us establish an analogy with the diffusion equation: We saw that we can derive the diffusion equation from the continuity equation supplemented with Fick's law, $\vec{j} = -D\nabla n$. In this example, the current is proportional to the pressure gradient, hence $\vec{j} \propto -\nabla p$. The density of the water is constant, and the height of the water column is what compensates for the non-uniform currents. The height of the column is also proportional to the pressure (up to an additive constant, which is inconsequential since only pressure *gradients* matter here). We therefore find that (in the absence of pumping) the height obeys the diffusion equation in 2D,

$$\frac{\partial h}{\partial t} = D\nabla^2 h, \tag{3.82}$$

where D is a constant (not to be confused with actual diffusion! We nevertheless use this notation to emphasize the formal analogy). If we don't pump water into or out of the well, its level will equilibrate. What happens if we continuously pump water out of the reservoir? In this case once the pumping is "turned on" we need to add a term proportional to $\delta(\vec{r})$ – a 2D δ-function – to the RHS of Eq. (3.82). Previously, we have found the solution of the diffusion equation (in any dimension) starting from δ-function initial conditions. Another way to think of this solution is as the solution of the equation

$$\frac{\partial h}{\partial t} = D\nabla^2 h + \delta(\vec{r})\delta(t), \tag{3.83}$$

albeit with initial conditions where $h = 0$ everywhere. Our problem (with continuous pumping) is similar to Eq. (3.83), but with the $\delta(t)$ replaced by a term proportional to the heaviside function $\theta(t)$, vanishing at negative times and taking the value 1 at positive values. This is a good illustration of the utility of Green's functions: We can now express the solution to our problem as an integral over the solution of Eq. (3.83) (can you see why?). This can be explicitly written in terms of Green's function of two-dimensional diffusion, as

$$H(r,T) \propto -\int_0^T \frac{e^{-\frac{r^2}{4Dt'}}}{4\pi Dt'} dt', \tag{3.84}$$

where the minus sign accounts for the fact that we are pumping water *out* of the well. To proceed, let us define $u \equiv \frac{r^2}{4Dt'}$ leading to

$$H(r,T) \propto -\frac{1}{4\pi D}\int_{\frac{r^2}{4DT}}^{\infty} \frac{e^{-u}}{u} du. \tag{3.85}$$

This is sometimes referred to as the "well function" and can be expressed using the *exponential integral function*:

$$E_1(x) = \int_x^{\infty} \frac{e^{-u}}{u} du. \tag{3.86}$$

For small, positive arguments, this function diverges logarithmically. This can be seen by integrating by parts:

$$E_1(x) = \log(u)e^{-u}|_x^{\infty} + \int_x^{\infty} e^{-u}\log(u)du. \tag{3.87}$$

The integral in Eq. (3.87) converges also when we set its lower limit as 0, and the result is known as the Euler–Mascheroni constant, γ_E. Thus we find that

$$E_1(x) \approx -\gamma_E - \log(x). \tag{3.88}$$

This function in fact pops up in numerous applications in physics and maths (and we will see this integral again later on in Chapter 6).

So close to the well (how close?) the height obeys

$$H \propto \gamma_E + \log\left(\frac{r^2}{4DT}\right). \tag{3.89}$$

Fig. 3.4(a), taken from Theis (1935), shows the excellent agreement of this prediction with the measurements. In fact, this is mathematically analogous to a recent application in the context of quantum optics, see Pugatch *et al.* (2014) and Fig. 3.4(b).

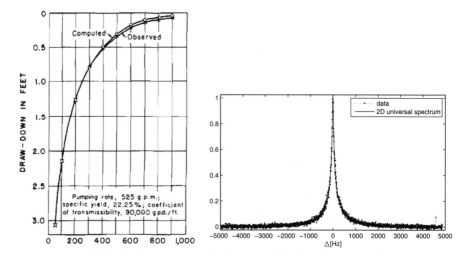

Figure 3.4 (left) The "well function" of Eq. (3.85), and the measurements of water height near a well during constant pumping. From Theis (1935). (right) The same mathematics applies in the context of quantum optics. Adapted Figure 1 with permission from Pugatch *et al.* (2014). Copyright 2014 by the American Physical Society.

3.4   Summary

In this chapter, we first generalized the random walker of Chapter 2 to the case where a force can act on the particle, thereby biasing the random motion in a space-dependent manner – described by the Langevin equation. We introduced an important discretization procedure that allows us to simulate such equations, and is also helpful conceptually. We next derived a PDE describing the time evolution of the probability distribution, the Fokker–Planck equation. In the case of particles diffusing in a potential due to thermal fluctuations, we derived a relation between the diffusion constant and the mobility of the particles, known as the Einstein relation. We next illustrated the philosophy behind the Langevin equation by discussing a discrete version of it, recently utilized in the context of cell size regulation. Similarly, we derived a PDE for the temporal evolution of the prices of options – the Black–Scholes equation – which resembled the Fokker–Planck equation of a freely diffusing particle. Finally, we showed how a similar diffusion equation arises in a very different context – pumping water out of a well – and how the Green function we derived in the context of the random walk problem can solve this problem and lead to interesting physical insights.

For further reading Gardiner (2009) discusses Langevin and Fokker–Planck in a far more comprehensive way than that outlined in this chapter, including numerous additional applications. Paul and Baschnagel (2013) has an extended discussion of applications to finance going well beyond the discussion in this chapter, as well as applications to quantum mechanics. Bressloff (2014) provides numerous additional examples involving stochasticity in the context of cell biology. Schuss (2009) and Honerkamp (1993) contain many additional physics applications.

3.5   Exercises

3.1   Dephasing in Nuclear Magnetic Resonance (NMR)

A nuclear spin in a constant magnetic field B is described by the Hamiltonian $H = -\begin{pmatrix} \omega_0 & 0 \\ 0 & -\omega_0 \end{pmatrix}$ where $\omega_0 \propto B$ is half the Larmor frequency, and we set $\hbar = 1$. The Schrödinger equation $i\dot{\Psi} = H\Psi$ for the time evolution of wavefunction $\Psi = c_1 |\uparrow\rangle + c_2 |\downarrow\rangle = \begin{pmatrix} c_1 \\ c_2 \end{pmatrix}$ yields the differential equations $\dot{c}_1 = i\omega_0 c_1$, $\dot{c}_2 = -i\omega_0 c_2$.

To account for fluctuations δB of the magnetic field due to coupling with other spins, the differential equations are modified by a noise term $\delta\omega$: $\dot{c}_1 = i(\omega_0 + \delta\omega)c_1$, $\dot{c}_2 = -i(\omega_0 + \delta\omega)c_2$ and the solutions by a phase factor: $c_1 = c_1(t_0)\,e^{i(\omega_0 t + \varphi(t))}$, $c_2 = c_2(t_0)\,e^{-i(\omega_0 t + \varphi(t))}$. The modified differential equations and solutions equation yield a Langevin equation for the time evolution of the phase: $\dot{\varphi} = \delta\omega$ (that is, $\dot{\varphi} \propto \delta B$).

(a) Given the Langevin equation $\frac{d\varphi}{dt} = \xi$ where ξ is a Gaussian random variable representing fluctuations of the magnetic field with autocorrelation function $C(t) = W^2 e^{-t/\tau}$, calculate $\langle e^{i\varphi(t)} e^{-i\varphi(t')} \rangle$ which determines the spin relaxation.

   Hint: You may use the relation $\langle e^X \rangle = e^{\langle X^2 \rangle/2}$, which holds for a normal random variable X.

(b) What happens for short and long correlation times of the noise?

See also Amir, A., Lahini, Y., and Perets, H. B. Classical diffusion of a quantum particle in a noisy environment, *Physical Review E*, 79(5), p.050105 (2009) for another application building on these results.

3.2   Boltzmann Distribution

Show that the Boltzmann distribution is a stationary solution of the Fokker–Planck equation for a general $V(x)$, $D(x)$, and $\mu(x)$ under the assumption of the Einstein relation between D and μ.

3.3   Fokker–Planck Equation for an Ornstein–Uhlenbeck Process

Consider the Langevin equation

$$\frac{dv}{dt} = -\frac{v}{\tau} + \xi$$

with ξ a white noise $\langle \xi(t)\,\xi(t') \rangle = A\delta(t - t')$.

(a) What is the corresponding Fokker–Planck equation?

(b) If $v(t = 0) = v_0$, what is the velocity distribution at time t? *Hint:* you may find it useful to represent v at time t given a noise $\xi(t')$, and rely on the statistics of the noise to infer the distribution.

(c) What happens as $t \to \infty$?

(d) What is the thermodynamic interpretation of (c)? How does A depend on temperature?

3.4   Particle in a Potential

Consider a particle with mass m in a 1d potential $V(x)$ that interacts with a thermal bath. The Langevin equation for its position $x(t)$ is

$$m\frac{d^2x}{dt^2} + \gamma\frac{dx}{dt} = -V'(x) + \zeta(t), \text{ where } \zeta(t) = m\xi(t) \text{ and}$$

$$\xi \text{ is white noise with } \langle\xi(t)\xi(t')\rangle = \Gamma\delta(t-t').$$

(a) Write down the Fokker–Planck equation for the joint distribution $p(x,v,t)$ of the position x and velocity v at time t.
(b) Find the stationary probability distribution. How is Γ related to γ and the temperature T?

In the rest of this problem, we specialize to a harmonic potential $V(x) = \frac{1}{2}kx^2$.

Let us denote by $p_X(x,t)$ and $p_V(v,t)$ the (marginal) distributions of the position and velocity at time t, respectively,

$$p_X(x) = \int p(x,v,t)dv, \text{ and } p_V(v) = \int p(x,v,t)dx.$$

(c) Find the distribution $p_X(x,t)$ of position at time t. Check that in the limit $t \to \infty$, the result is consistent with what you would expect from part (b).
(d) Find the distribution $p_V(v,t)$ of velocity at time t.
(e)* Find the joint distribution $p(x,v,t)$.

We thank Pétur Rafn Bryde for help in creating this problem.

3.5   Diffusion in a Logarithmic Potential

Consider diffusion of a particle in a logarithmic potential $V(x) = a\log(|x|) + U(x)$, where the correction $U(x)$ is negligible for large $|x|$ but at small $|x|$ it ensures that $V(x)$ does not diverge. Assume that the temperature is such that $kT = 1$.

(a) Write the Langevin equation and the corresponding Fokker–Planck equation for diffusion in a logarithmic potential, assuming that we are in the overdamped limit and that the diffusion constant is D.
(b) For $a > 1$, what is the steady-state solution (away from the origin)? What is the issue for $a < 1$?
(c) For $a < 1$, guess a scaling solution for the dynamics of the probability distribution: $p(x,t) = 1/t^b f(x/t^c)$. Find the value of c for which f obeys a second-order *ordinary* differential equation. Find b from demanding that the probability distribution is normalized.
(d) Write the ODE that f obeys. What are the boundary conditions?

We thank Ori Hirschberg for help in creating this problem.

3.6 Cell Size Control

(a) Simulate the following discrete map in MATLAB or the software of your choice:

$$x_{n+1} = (1 - \alpha)x_n + \xi,$$

where ξ is a random variable drawn from a Gaussian distribution, with mean 0 and standard deviation of 0.1, and for values of α of 0, 0.1, 0.5, and 1.

(b) For each instance, run the map for 100,000 steps, and plot the distribution of x.

(c) Calculate numerically the Pearson correlation coefficient between the variables x_n and x_{n+1}.

Hint: the MATLAB commands randn and corrcoeff could be useful.

3.7 Cell Size Control*

In the notes, we derived the Pearson correlation coefficient (i) between the volume of the mother at birth and the volume of the daughter at birth, and (ii) between the volume at birth and the interdivision time (the time from birth to division).

(a) Simulate the model for $\alpha = 1/2$ for 10,000 division events (following a single-lineage), and computationally estimate the two aforementioned Pearson correlation coefficients.

(b) Analytically derive the Pearson correlation coefficients (iii) between the interdivision times of mothers and daughters, and (iv) between the interdivision times of two cousins. What are the correlations for arbitrary α, and for $\alpha = 1/2$? Compare your results to recent experiments on human HeLa cells [Sandler *et al.* *Nature* (2015)] that obtained a Pearson correlation coefficient ≈ 0.04 for mother–daughter interdivision times, ≈ 0.63 for cousin–cousin interdivision times, and ≈ 0.76 for sister–sister interdivision times. Are the model and experiment consistent?

3.8 Black-Scholes When Stocks Do a Normal Random Walk

(a) For the case of a *multiplicative* random walk (considered in this chapter), if the volatility is $\sigma = \frac{1}{\sqrt{day}}$, and the current value of a stock is 10\$, how much would a "call option" be worth to buy that stock at a price of 5\$ in 1, 5, and 10 days from now? Assume that the interest rate $r = 0$.

(b) Make the (unrealistic) assumption that the stock market does a random walk rather than a *multiplicative* random walk. Repeat the derivation of the Black–Scholes equation. What is the (discrete) backwards recursive relation?

(c) By Taylor expanding the equation to the appropriate order, derive the PDE corresponding to the discrete equation you derived in part (b).

3.9 Price of Options within Black–Scholes

Prove that the option price increases as the time difference between now (t_0) and the exercising time (t_1, $t_1 > t_0$) increases, for a given strike price at t_1 and stock price S, and under the assumption that the interest rate is positive.

3.10   Divisibility**

Prove that a random variable uniformly distributed in $[0, 1]$ cannot be written as the sum of two independent and identically distributed (i.i.d.) random variables.

3.11   Active Matter

A bacterium moves in 1D at a velocity with spatial-dependent magnitude $v(x)$, and a constant rate r of reversing its direction of motion. This type of dynamics is known as "run and tumble."

(a) Set up the Fokker–Planck equation for the bacterium. *Hint:* Make sure that the probability is conserved, e.g., by making sure that certain terms in the FP equation can be expressed as divergences (i.e., spatial derivative in this 1D problem).
(b) Find how the bacterial density depends on the position in steady-state. Can you provide some intuition for the result you obtained?

3.12   Population Dynamics

A culture consists of precisely $N \gg 1$ bacterial cells, each of which belongs to either species A or B. At every (discrete) time step, one of the N individuals, chosen randomly and uniformly, reproduces, after which one of the $N + 1$ individuals is removed from the culture, also chosen randomly and uniformly.

(a) Write the probabilities for the number of cells of species A to (i) remain unchanged. (ii) increase by one. (iii) decrease by one.
(b) Use the results to write a Fokker–Planck equation for the dynamics of $P(x, t)$, where x is the number of cells of type A. *Hint:* attempt to identify terms of the structure $-2g(x)p(x, t) + g(x + 1)p(x + 1, t) + g(x - 1)p(x - 1, t)$, and take the continuous limit of this expression (for large N).
(c) What happens at very long times?
(d) Use the results of (b) to estimate the scaling (with N) of the duration it takes the asymptotic behavior of (c) to kick in, starting from an initial condition where both species are equally represented in the population.
(e) Verify the scaling of (d) using numerical simulations.

Jiseon Min helped create this problem.

4   Escape Over a Barrier

Now transmuted, we swiftly escape as Nature escapes

(Walt Whitman, *Leaves of Grass*)

An important application of the Fokker–Planck equation is in solving the "escape-over-a-barrier" problem. In many cases, a lower energy state of the system exists, but in order to achieve it, the system has to overcome an "activation barrier," the magnitude of which we will denote by U (see Fig. 4.1). If the barrier is high compared to the scale of kT, then only a rare fluctuation will manage to get the system across the barrier. In the case of chemical reactions, this rate will determine the reaction rate. Clearly, many degrees of freedom may be involved in such scenarios, though often chemists speak of a "reaction coordinate(s)," effectively reducing the problem to that of a lower dimensionality. Indeed, in this chapter we will solve the problem of escape from a metastable state in a general dimension, and see that the results are similar to those of the 1D problem. There are many other contexts in physics in which this problem arises: to name but a few, these include superconductivity and superfluidity (Ambegaokar *et al.* 1980), dislocation theory (Bruinsma, Halperin, and Zippelius 1982; Amir, Paulose, and Nelson 2013b), and biophysical processes (Möbius, Neher, and Gerland 2006; Suzuki and Dudko 2010). See Coffey, Kalmykov, and Waldron (2004) and Gardiner (2009) for numerous additional examples.

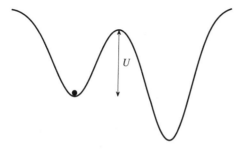

Figure 4.1   A one-dimensional escape-over-a-barrier problem.

Nucleation There are many scenarios where the formation of a phase of matter involves nucleation (e.g., think of vapor bubbles in a liquid) and the kinetics of the problem maps to an "escape-over-a-barrier" problem: The free energy of the new phase is lower than that of the initial phase, but there is a free energy barrier to overcome. Similarly, Problem 4.2 deals with the nucleation of a pair of vortices in a superfluid (or superconductor) – formation of the pair of vortices is energetically favorable when they are far apart, but there is a barrier to overcome when they are too close. This is an example of a nucleation process occurring, effectively, in a two-dimensional space.

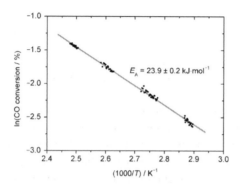

Figure 4.2 Measured rate of a chemical reaction (CO conversion) as a function of temperature, manifesting the Arrhenius relation of Eq. (4.1). Adapted from Roos *et al.* (2011).

In the 1880s, Svante Arrhenius found, experimentally, that the rate of certain chemical reactions can be well described by a temperature-dependence of

$$\Gamma \propto e^{-U/kT}. \tag{4.1}$$

See, for example, Fig. 4.2 showing the temperature-dependent-rate of a chemical reaction.

Noting the similarity to the Boltzmann distribution, this is plausible, but what determines the prefactor? Only in the 1940s did Kramers solve a one-dimensional "escape-over-a-barrier" problem, calculating the escape rate of a particle from the potential depicted in Fig. 4.1.

Let us define $\beta \equiv \gamma/m$, the ratio of the viscosity (defined as the inverse of the mobility, $\gamma \equiv 1/\mu$) to the mass (which has dimensions of frequency). In the over-damped case, Kramers found that

$$\Gamma = \frac{\omega_A \omega_S}{2\pi\beta} e^{-U/kT}, \tag{4.2}$$

where ω_A, and ω_S are the angular frequencies of the particle in the harmonic potential at the bottom of the well and the (inverted) harmonic potential at the top of the barrier (in other words, if we approximate the potential near the minimum or maximum as $\frac{1}{2}kx^2$, $\omega = \sqrt{\frac{k}{m}}$). Note that in the overdamped limit, the mass should not appear. Indeed, the dependence of the frequencies on mass cancels with it. Also, note that the fact that the Einstein relation (between mobility and diffusion constants) elucidates why the rate of going over the barrier via thermal fluctuations is governed by the mobility.

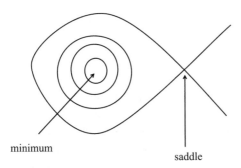

minimum

saddle

Figure 4.3 A two-dimensional "escape-over-a-barrier" problem.

We will prove this result, as a particular case of a more general theory developed by James Langer (1969). Consider, for example, an "escape-over-a-barrier" problem in 2D. Fig. 4.3 shows the equipotential contours of the potential energy landscape, which should be familiar to those with a background in navigation (as it is essentially a "topographic map" of the potential landscape).

For large barriers compared to the energy scale kT, the most likely way that a thermal fluctuation would make the particle get out of the basin of the minimum is via the saddle-point (again, this is intuitive if you think of the best way to cross a mountain range – the mountain pass is typically a saddle!). We will calculate the rate of particles across the saddle-point in the overdamped limit, and show that it can be succinctly written in terms of the *Hessians* of the potential at the minimum and the saddle. Recall that the $N \times N$ Hessian matrix of a function $f(x_1, x_2..x_N)$ (at a given point) is defined as

$$H_{ij} = \frac{\partial^2 f}{\partial x_i \partial x_j} \tag{4.3}$$

(see also Appendix D). Moreover, we know that all eigenvalues of H will be positive at the minimum, while at the saddle we have at least one positive and one negative one.

The Langer formula states that

$$\Gamma = \frac{|\lambda|}{2\pi} \sqrt{\frac{|\det H^m|}{|\det H^s|}} e^{-U/kT}. \tag{4.4}$$

Here, H^m and H^s are the *Hessian* matrices of the potential around the minimum and saddle, respectively, and λ is the negative eigenvalue of the matrix $H^s M$ (assumed to be unique), where M is the mobility matrix. Before going through the rather lengthy derivation of this formula, let us see how it reduces to Kramers, problem in 1D, and also find the result in the case where inertia matters in 1D – which as we shall see can be thought of as a 2D "escape-over-a-barrier" problem.

For the 1D problem, the potential near the minimum can be approximated as

$$V(x) \approx \frac{1}{2}k_A x^2, \tag{4.5}$$

hence the Hessian is a 1×1 matrix whose entry is the second derivative of V at that point, i.e., $k_A = m\omega_A^2$. Similarly, the Hessian at the saddle is $k_S = -m\omega_S^2$. In this case $\lambda = k_S \mu$, where $\mu = 1/\gamma$ is the mobility of the particle. Plugging this into Eq. (4.4), we recover Eq. (4.2). Problem 4.5 solves the Fokker–Planck equation *numerically* to directly compute the escape rate for 1D overdamped motion, recovering the results of the above formula.

4.1   Setting Up the Escape-Over-a-Barrier Problem

The Fokker–Planck equation that we will be working with is

$$\frac{\partial p}{\partial t} = \sum \frac{\partial}{\partial \eta_i} \left(M_{ij} \left[\frac{\partial V}{\partial \eta_j} + kT \frac{\partial}{\partial \eta_j} \right] p \right). \tag{4.6}$$

Here M_{ij} is the mobility matrix, which multiplies the vector to its right, and the derivative $\frac{\partial}{\partial \eta_i}$ operates on both of the terms on the RHS. This equation is a special case of Eq. (3.9) where we used the Einstein relation $D = kTM$: We previously saw that this holds in the one-dimensional overdamped case, and the same derivation can easily be generalized to the overdamped scenario in any spatial dimension (where now the relation will be between two *matrices*).

For a constant mobility matrix one may move the term M_{ij} outside the derivative. We will not discuss this here, but in various cases the above equation accounts for a spatially dependent mobility. The first term on the RHS is the contribution of the force to the current, while the second one is due to the diffusive term. The Langevin equation corresponding to Eq. (4.6) is

$$\frac{d\vec{\eta}}{dt} = M[-\nabla V] + \vec{\xi}, \tag{4.7}$$

with $\langle \xi_i \xi_j \rangle = F_{ij}\delta(t - t')$.

What is the relation between F, M and the diffusion matrix D? Consider first the case of no potential. We have seen before in *one dimension* that this leads to a diffusion equation, with

$$D = F/2. \tag{4.8}$$

However, since F is symmetric, we can always diagonalize it, and in the new basis every coordinate will perform independent diffusion – hence this relation is completely general.

4.2 Application to the 1D Escape Problem

Consider the Langevin equation of a particle *with* inertia in one dimension. We have seen that this can be written as

$$\begin{pmatrix} \dot{x} \\ \dot{p} \end{pmatrix} = -M \begin{pmatrix} \frac{\partial \tilde{V}}{\partial x} \\ \frac{\partial \tilde{V}}{\partial p} \end{pmatrix} + \vec{\xi}, \tag{4.9}$$

with $\tilde{V} = \frac{p^2}{2m} + V(x)$, and $\xi = \begin{pmatrix} 0 \\ \zeta \end{pmatrix}$ and $M = \begin{pmatrix} 0 & -1 \\ 1 & \gamma \end{pmatrix}$.

Beware of the notation. Note that in this section p is the particle momentum rather than the probability distribution, which we will denote by P to be distinct from it!

Also note that this probability distribution depends on both components of the vector $\vec{x}$, namely x and p. The noise is "white," i.e., $\langle \xi_i \xi_j \rangle = F_{ij}\delta(t - t')$ (we will cover other forms of noise in Chapter 5), and we previously showed it has the structure

$$F = \begin{pmatrix} 0 & 0 \\ 0 & C \end{pmatrix}. \tag{4.10}$$

This leads to

$$\frac{\partial P(\vec{x})}{\partial t} = -\nabla \cdot (\vec{J}), \tag{4.11}$$

with

$$\vec{J} = -M(\nabla \tilde{V})P - D\nabla P, \tag{4.12}$$

where we have recognized the first term with the (generalized) forces and the second term as "Fick's law," coming from the noise term. As we have argued in the previous section, the latter term can be expressed in terms of the noise as

$$D_{ij} = F_{ij}/2, \tag{4.13}$$

hence only one of the components of the matrix D is non-vanishing, which we shall term D.

Let us find the corresponding Fokker–Planck equation. The current $\vec{J}_f$ due to the force is

$$P(x, p)M \begin{pmatrix} f \\ -v \end{pmatrix} = P \begin{pmatrix} v \\ -V'(x) - \gamma v \end{pmatrix}. \tag{4.14}$$

Hence its divergence is

$$\nabla \cdot \vec{J}_f = v\frac{\partial P}{\partial x} - V'(x)\frac{\partial P}{\partial p} - \frac{\gamma}{m}\frac{\partial (Pp)}{\partial p}. \tag{4.15}$$

The current $\vec{J}_d$ due to "Fick's law" is $-D\nabla P = \begin{pmatrix} 0 \\ -D\frac{\partial P}{\partial p} \end{pmatrix}$, with $D = C/2$ yet to be determined. Therefore, its divergence is

$$\nabla \cdot \vec{J}_d = -D\frac{\partial^2 P}{\partial p^2}, \tag{4.16}$$

hence the Fokker–Planck equation reads

$$\frac{\partial P}{\partial t} + v\frac{\partial P}{\partial x} - V'(x)\frac{\partial P}{\partial p} - \frac{\gamma}{m}\frac{\partial(Pp)}{\partial p} - D\frac{\partial^2 P}{\partial p^2} = 0. \tag{4.17}$$

This is known as the Klein–Kramers equation.

As in Chapter 3, we can now determine D by imposing that the Boltzmann distribution should be a stationary solution of this equation. For simplicity, let's take $V = 0$. Then we have $P \propto e^{-\frac{p^2}{2mkT}}$, and

$$-\frac{\gamma}{m}\frac{\partial(Pp)}{\partial p} - D\frac{\partial^2 P}{\partial p^2} = 0. \tag{4.18}$$

Therefore,

$$-\frac{\gamma}{m}Pp - DP' = const \tag{4.19}$$

$\Longrightarrow$

$$-\frac{\gamma}{m}e^{-\frac{p^2}{2mkT}}p + De^{-\frac{p^2}{2mkT}}\frac{2p}{2mkT} = const, \tag{4.20}$$

hence

$$D = \gamma kT. \tag{4.21}$$

This is another example of a fluctuation–dissipation relation – like the Einstein relation – this time relating the diffusion of *velocity* to the damping. But what about the fate of Eq. (4.6) in this case? Clearly, here the matrices D and M are not proportional. Yet it turns out that Eq. (4.6) is still correct! This is due to the antisymmetric nature of the off-diagonal part of M, and the relation of Eq. (4.21) applied to the single non-vanishing entry of D. Indeed, the two terms

$$M_{ij}\frac{\partial^2 P}{\partial \eta_i \partial \eta_j}; M_{ji}\frac{\partial^2 P}{\partial \eta_j \partial \eta_i}, \tag{4.22}$$

cancel due to the equality of mixed partials and the structure of M.

This implies that the result we are about to derive, Eq. (4.4), is still applicable also for this case.

Escape rate vanishes for large and small damping Problem 1 utilizes the Langer formalism in this context to derive an escape rate for a particle with inertia. Not surprisingly, for sufficiently strong damping the results of Eq. (4.2) are recovered. As $\beta \to \infty$, the environment becomes so viscous the particle cannot get over the barrier (i.e., the prefactor of the Arrhenius dependence of Eq. (4.1) vanishes). What happens at small damping, $\beta \to 0$? In this case, according to the fluctuation–dissipation theorem derived above, the diffusion constant vanishes (note that β and γ are proportional to each other). This implies that for small damping (small viscosity), there will be no fluctuating forces to drive the particle over the barrier, and the rate of escape should vanish too! However, the equation derived for the case of a particle with inertia (using the Langer formalism) does not vanish as $\beta \to 0$. This is a shortcoming of the derivation: as we shall see, the derivation will involve the assumption of a Boltzmann distribution in the vicinity of the metastable state. This quasi-equilibrium cannot be achieved as $\beta \to 0$ since in this case the environment cannot thermalize the diffusing particle sufficiently fast. For further details on how the Fokker–Planck can be solved in this case, see Coffey, Kalmykov, and Waldron (2004).

4.3   Deriving Langer's Formula for Escape-Over-a-Barrier in Any Spatial Dimension

Starting with Eq. (4.6), we will make the ansatz

$$p = p_{eq}\xi, \tag{4.23}$$

with $p_{eq} = e^{-V/kT}$ the equilibrium distribution (up to a normalization constant). Since the Boltzmann distribution solves Eq. (4.6), why should we bother to look for such an ansatz? The answer is the initial condition: If beyond the large barrier lies an infinitely deep well, the weight of the Boltzmann distribution in the metastable state is vanishingly small. However, the question we ask is different: *Given* that we start in that state (which can come about for a multitude of reasons), what is the rate of leakage out of it? In that case, thermal equilibration *within* the region around the local minimum will occur relatively fast, but the deviations from it around the saddle will, as we shall see, lead to the finite but slow leakage out-of-it, which would ultimately lead to the establishment of a true equilibrium with vanishing probability to be in the metastable state.

Plugging the ansatz into Eq. (4.6), we find that stationarity of the solution implies that

$$\sum_{i,j} M_{ij}\left[p_{eq}\frac{\partial^2 \xi}{\partial \eta_i \partial \eta_j} - \frac{1}{kT}p_{eq}\frac{\partial V}{\partial \eta_i}\frac{\partial \xi}{\partial \eta_j}\right] = 0. \tag{4.24}$$

Which can also be written in the compact notation:

$$\sum_{i,j}\frac{\partial}{\partial \eta_i}M_{ij}p_{eq}\frac{\partial \xi}{\partial \eta_j} = 0. \tag{4.25}$$

Let us expand the potential energy around the saddle-point. Denoting the deviation from the saddle-point by $\delta\vec{\eta}$ (not to be confused with an infinitesimal!) we have

$$V(\delta\vec{\eta}) \approx V_s + \sum_{i,j} \frac{1}{2} H^s_{ij} \delta\eta_i \delta\eta_j. \tag{4.26}$$

Leading to

$$\frac{\partial V}{\partial \eta_i} = \sum_k H^s_{ik} \delta\eta_k \tag{4.27}$$

(note that precisely at the saddle-point the derivatives vanish, and here the derivative is calculated in the vicinity of, but not precisely at, the saddle-point). Plugging this into the Fokker–Planck equation we find that

$$\sum_{i,j} M_{ij} \left[\frac{\partial^2 \xi}{\partial\eta_i \partial\eta_j} - \frac{1}{kT} \sum_k H^s_{ik} \delta\eta_k \frac{\partial \xi}{\partial\eta_j} \right] = 0. \tag{4.28}$$

Now comes a second ansatz: We will assume that the function ξ depends on a single-parameter which is a linear combination of the components of $\delta\vec{\eta}$. This will define the "direction" through which we will be crossing the "mountain-range" (saddle), which does not have to be perpendicular to it since the mobility tensor might bias us to prefer a diagonal crossing. Therefore we are seeking

$$\xi = \xi(u); u = u_1 \delta\eta_1 + u_2 \delta\eta_2 \ldots \tag{4.29}$$

$$\Longrightarrow$$

$$\frac{\partial \xi}{\partial\eta_j} = \xi' \frac{\partial u}{\partial\eta_j} = \xi' u_j, \tag{4.30}$$

$$\frac{\partial^2 \xi}{\partial\eta_i \eta_j} = \xi'' u_i u_j. \tag{4.31}$$

Plugging this into Eq. (4.28), we find that

$$\sum_{i,j} M_{ij} \left[kT \xi'' u_i u_j - \sum_k H^s_{ik} \delta\eta_k \xi' u_j \right] = 0. \tag{4.32}$$

To solve this, we are going to assume that $\xi(u)$ takes a particular form (Eq. (4.39) below). However, in order to make this form less mysterious, we will obtain the functional form $\xi(u)$ *must* have by considering the partial differential equation of Eq. (4.32) along one of the axes, where it becomes a simple ordinary differential equation.

Consider the equation along one of the axes, k, and let us make a change of variable $x \equiv u_k \delta\eta_k$. According to this choice, along this axis $\xi(u) = \xi(x)$, and the equation has the structure

$$C_1 \xi''(x) + C_2 x \xi'(x) = 0, \tag{4.33}$$

with

$$C_1 = kT \sum_{i,j} M_{ij} u_i u_j \tag{4.34}$$

and

$$C_2 = -\sum_{i,j} M_{ij} H^s_{ik} u_j. \tag{4.35}$$

This is easy to solve; defining $g = \xi'$ leads to

$$C_1 g' + C_2 x g = 0 \implies [\log(g)]' = -x C_2 / C_1. \tag{4.36}$$

Hence

$$g = A e^{-\frac{C_2 x^2}{2C_1}} \tag{4.37}$$

and

$$\xi(x) = A \left[\int_{-\infty}^x e^{-\frac{C_2 x'^2}{2C_1}} dx' \right] + B \tag{4.38}$$

(which can of course be written in terms of the error function).

Since ξ depends on a single-parameter u, its form must be

$$\xi(u) = A \left[\int_{-\infty}^u e^{-Cu'^2} du' \right] + B. \tag{4.39}$$

At this stage we know that this must be the functional form of $\xi(u)$, but we still don't know whether our ansatz will solve Eq. (4.32). Plugging this into Eq. (4.32) we find that

$$\sum_{i,j} M_{ij} \left[kT e^{-Cu^2} \left(-2C \sum_k u_k \delta\eta_k \right) u_i u_j - \sum_k H^s_{ik} \delta\eta_k e^{-Cu^2} u_j \right] = 0. \tag{4.40}$$

After eliminating the common Gaussian factor, all terms are linear in the elements $\delta\eta_i$, and the coefficient of each $\delta\eta_i$ must therefore vanish independently. Therefore, a necessary and sufficient condition for our ansatz to work is that for every k we have

$$\sum_{i,j} M_{ij} [kT(-2Cu_k) u_i u_j - H^s_{ik} u_j] = 0. \tag{4.41}$$

The last term can be rewritten as

$$\sum_{i,j} H^s_{ki} M_{ij} u_j = [H^s M \vec{u}]_k, \tag{4.42}$$

where we relied on the fact that $[H^s]^T = H^s$ since it is a Hessian.

The first term can be rewritten as $[-2CkT \sum_{i,j} M_{ij} u_i u_j] u_k$. Therefore, the ansatz will indeed solve the Fokker–Plack equation, as long as $\vec{u}$ is an eigenvector of the matrix $H^s M$, with eigenvalue

$$\lambda = -2CkT \sum_{i,j} M_{ij} u_i u_j. \tag{4.43}$$

Looking at Eq. (4.38), clearly B will not matter for the rate across the saddle – since it is associated with the Boltzmann distribution (with no correction), which is a stationary solution of the FP equation. We are assuming that the escape rate is low (i.e., that the temperature is sufficiently low), and that we started in the metastable state. Hence we should have negligible weight past the barrier, implying that $B = 0$. Since our assumption is that at time $t = 0$ the particle is in the metastable state, the direction of positive u must be going from the saddle *towards* the metastable state, such that the integral of Eq. (4.38) will approach a positive constant away from the saddle on the side of the metastable state, and will vanish on the other side of the saddle. The value of A is then set by normalization of the probability distribution, i.e., $A = 1/Z$ with Z the partition function, that we will shortly evaluate (see Appendix D for a reminder regarding partition functions).

The "escape-over-a-barrier" problem is a good example of timescale separation: We are assuming that the temperature is low (compared with the barrier height), such that the rate of "leaking" into the global minimum is small. Thus thermal equilibrium is achieved in the region of the metastable state, and the probability distribution is mostly concentrated in that region for the timescales we are interested in. Note that if we wait long enough *true* thermal equilibrium will be achieved everywhere: In that case the probability will be concentrated in the true minimum rather than the metastable state (formally, that corresponds to $B = 1$ and $A = 0$). However, our transient solution is an excellent approximation that allows us to accurately calculate the probability *flux* across the saddle-point. The timescale associated with crossing the barrier (the reciprocal of the rate of crossing) is assumed to be much longer than the timescale for establishing the Boltzmann distribution in the vicinity of the metastable state.

Note that the choice of C is arbitrary – since we can always scale the magnitude of $\vec{u}$ accordingly (i.e., if $C \to 2C$ we take $u_i \to u_i/\sqrt{2}$). For convenience, we choose $2CkT = 1$. Note that the sum $M_{ij} u_i u_j$ is positive since M is positive–definite (since the diagonal is positive and the off-diagonal anti-symmetric), hence the eigenvalue is *negative*. We will assume that only one eigenvalue of the matrix $H^s M$ is negative (one "unstable" direction).

Therefore, we have found a stationary solution which is a product of the Boltzmann distribution and the correction term ξ, which we found to depend on the distance along a certain direction from the saddle-point, which turned out to be an eigenvector of a particular matrix. Note that ξ vanishes away from the saddle on one side of it – clearly the one beyond the barrier – and is approximately constant far away on the other side, e.g., in the vicinity of the minimum. This is expected, since the solution near the minimum is proportional to the Boltzmann distribution. To proceed, let us calculate the partition function Z, which determines A in Eq. (4.39) through the normalization relation $A = 1/Z$, which reads

$$Z = \int p_{eq}(\vec{\eta})\xi(\vec{\eta})d\eta_1 d\eta_2 \ldots d\eta_n. \tag{4.44}$$

The probability distribution is dominated by the region surrounding the minimum. Without loss of generality we can set the potential to be zero at the minimum. Denoting the Hessian around the minimum by H^m we have

$$Z \approx \xi(minimum) \int e^{-\sum_{i,j} \frac{H^m_{ij}}{2kT} \delta x_i \delta x_j} d\delta x_1 d\delta x_2 \ldots \delta x_n, \tag{4.45}$$

with $\vec{\delta x}$ the vector of deviations from the minimum, and $\xi(minimum)$ approximately constant and equal to $\sqrt{\pi/C} = \sqrt{2\pi kT}$. This is a standard Gaussian integration (see Appendix B for details). We can change basis to one where H^m is diagonal, using a *unitary* matrix (since H^m is symmetric). Since the transformation is unitary its Jacobian is 1, and we find that

$$Z \approx \sqrt{2\pi kT} \sqrt{\left(\frac{2\pi kT}{\lambda_1}\right)\left(\frac{2\pi kT}{\lambda_2}\right) \cdots \left(\frac{2\pi kT}{\lambda_n}\right)}, \tag{4.46}$$

hence

$$Z \approx \sqrt{2\pi kT} \frac{(2\pi kT)^{n/2}}{(det[H^m])^{1/2}}. \tag{4.47}$$

Finally, we have to compute the current going through the saddle. Let us first find the current in the vicinity of the saddle. The formula for the current is

$$\vec{j} = -kTM\nabla p - M\nabla V p. \tag{4.48}$$

Since at the saddle-point $\nabla V = 0$, the second term will not contribute. Similarly, the term

$$-D\xi\nabla p_{eq} \approx M\xi p_{eq} \sum_{i,j} H^s_{ij}\delta\eta_j \tag{4.49}$$

will also be negligible. The contribution will come from the term

$$-\frac{kT}{Z}M p_{eq}\nabla\xi = -\frac{kT}{Z}M p_{eq}e^{-u^2/(2KT)}\vec{u}. \tag{4.50}$$

The direction of the flux is $-M\vec{u}$. As intuitively expected, if M is the identity matrix, the crossing of the saddle-point will be determined by the negative eigenmode of H^s.

It is convenient to perform this integration by calculating the flux through a plane perpendicular to $\vec{u}$ – since the term $e^{-u^2/(2kT)} = 1$ everywhere on this plane (passing through the saddle). To compute the flux through the surface, we have to take the scalar product of the current vector with the normal to the surface, $\vec{u}$. Since the current is pointing in the $M\vec{u}$ direction at every point in the vicinity of the saddle, this gives a constant factor of

$$\phi \equiv (M\vec{u}, \vec{u})/||\vec{u}|| = \sum_{i,j} M_{ij} u_i u_j /||\vec{u}|| = -\lambda/||\vec{u}||. \tag{4.51}$$

The overall current will be

$$I = kT\phi \int p_{eq} dv_1 dv_2 \ldots dv_{n-1}, \tag{4.52}$$

where the integration is done along the $n-1$ directions *perpendicular* to $\vec{u}$, and $p_{eq} = e^{-V/kT}/Z$. Note that this is strictly speaking *not* the equilibrium distribution, since although it has the Boltzmann distribution exponential dependence on the potential, its normalization (the term Z) was determined by the *metastable* state – the minimum we integrated over was the one the system is escaping from, not the true minimum.

In the vicinity of the saddle (which dominates the flux), the term p_{eq} can be approximated by

$$p_{eq} \approx e^{-U/kT} e^{-\sum_{i,j} \frac{H^s}{2kT} \delta\eta_i \delta\eta_j}. \tag{4.53}$$

Let us perform the calculation under the simplifying assumption of a matrix M equal to the identity matrix (for the general case, see Coffey, Garanin, and McCarthy 2001). In that case, $\vec{u}$ is an eigenvector of H^s, and hence the structure of H^s is

$$H^s = \begin{pmatrix} -\lambda & 0 & 0 & .. & 0 \\ 0 & & & & \\ 0 & H^s_{reduced} & & & \\ ... & & & & \\ 0 & & & & \end{pmatrix}. \tag{4.54}$$

We are interested in the integration in the *reduced subspace*. This is a standard Gaussian integration identical in form to that which we encountered in Eq. (4.45), and hence we have

$$\Gamma \approx e^{-U/kT} \frac{kT\phi}{Z} \frac{(2\pi kT)^{(n-1)/2}}{(|det[H^s_{reduced}]|)^{1/2}|}. \tag{4.55}$$

But according to the above form of H^s, we also have

$$|det[H^s_{reduced}]| = |det[H^s]/\lambda|. \tag{4.56}$$

In this case $\phi = \sqrt{|\lambda|}$ (since $||u|| = \sqrt{|\lambda|}$), hence the escape rate will be

$$\Gamma = e^{-U/kT} ||\vec{u}|| \frac{kT}{\sqrt{2\pi kT}\sqrt{2\pi kT}} \sqrt{|\lambda|} \sqrt{\left|\frac{det[H^m]}{det[H^s]}\right|}, \tag{4.57}$$

which gives us Eq. (4.4), since in this case $||\vec{u}||^2 = |\lambda|$.

4.4   Summary

This chapter discussed an explicit application of the Fokker–Planck equation, to a ubiquitous problem in physics, chemistry and biology – overcoming an energy barrier by thermal fluctuations. The dominant feature of the result is the celebrated Arrhenius dependence of the transition rate on the energy barrier between the metastable state and the saddle-point separating it from the true minimum. The solution illustrated the use of the Fokker–Planck equation in a potentially high dimension, and the Hessians of the potential energy at the metastable state and the saddle-point played key roles. Furthermore, it demonstrated the use of a separation of timescales, on which the physical ansatz used to solve the Fokker–Planck equation relied.

For further reading Coffey, Kalmykov, and Waldron (2004) provides a more elaborate discussion of the methods discussed in this chapter, using similar terminology and approach. For mathematically different approaches to the escape problem, see Gardiner (2009); Honerkamp (1993); Schuss (2009); Van Kampen (1992).

4.5   Exercises

4.1   1D Escape with Mass

Use Langer's formalism to find the rate of escape in 1D for a particle *with* inertia. How large should the damping be for your results to recover those of the overdamped case? Why?

4.2   Vortex and Dislocation Unbinding

It is known that the vortices in superconducting films interact via a logarithmic potential, i.e., the potential energy of two vortices a distance r apart is $V(r) = c \log(r/r_0)$, when $r \gg r_0$, with r_0 a microscopic distance cutoff below which the potential saturates at a constant. For concreteness, assume the potential has the precise form $v(r) = c \log[\sqrt{r_0^2 + r^2}]$. Note that this potential is *two-dimensional*.

In the presence of an external supercurrent, an additional term Bx will be added to the total energy, where x is the relative separation of the vortices along the x-direction. At a finite temperature, thermal fluctuations will lead to the nucleation of vortex pairs.

(a)   Show that in the presence of a magnetic field there is a saddle-point, and find its location and energy.

(b)   Using Langer's formalism, find the nucleation rate.

(c)   Two edge dislocations with opposite Burgers vectors in the x-direction have an anisotropic interaction $V(r) = c\left[\log(r/r_0) - x^2/r^2\right]$. In an external stress, there will be an additional component Ax. However, unlike vortices the dislocations can only move along the x-direction.
What is the nucleation rate in this case?

4.3 Sinai's Walk

Consider a one-dimensional random walk, characterized by a position U as a function of time x with stepsize δU and discrete time step δx. The output of one trajectory would be a function $U(x)$. For this problem, consider this function as a *potential energy*, in which **another** random walker will be moving. This is known as Sinai's walk.

Consider a particle random walking in the energy landscape $U(x)$, with step sizes δx. At every discrete time step δt, the random walker flips a coin and chooses a direction. If the energy $U(x)$ is lower in the direction chosen, the move is accepted. If it is higher, the move is accepted with probability $e^{-\Delta U}$, where ΔU is the energy difference (this corresponds to Arrhenius's law, where we have chosen here $kT = 1$ for convenience).

(a) Generate an energy landscape $U(x)$ by simulating a (single) random walker for $0 < x < 10^5$. Now find $\langle x^2(t) \rangle$ for a random walker following the dynamics explained above on the landscape $U(x)$, where $\langle \rangle$ denotes averaging over 10^4 or more runs (for a fixed landscape). Show an example of your numerics for the following parameters: $\delta U = 2$, $\delta x = 1$.

(b) Repeat (a) for 300 or more instances of the potential $U(x)$, and plot $\langle\langle x^2(t) \rangle\rangle$ for $0 < t < 10^5$, where $\langle\langle x^2(t) \rangle\rangle$ denotes an average of $\langle x^2(t) \rangle$ over different realizations of the potential. For each realization of $U(x)$, use an average over 100 realizations of $x(t)$ to calculate $\langle x^2(t) \rangle$.

(c) How does $\langle\langle x^2(t) \rangle\rangle$ behave for large t? It may help to plot $\langle\langle x^2(t) \rangle\rangle$ as a function of the appropriate power of $\log(t)$. Explain, qualitatively, where this scaling arises from.

4.4 Langer's Formula**

Prove Langer's formula without assuming M is proportional to the identity matrix.

4.5 Simulating Escape Over a Barrier

(a) Consider the following potential in the 1D overdamped limit:

$$U(x) = x^2/10 - e^{-x^2} - 5e^{-(x-3)^2}.$$

Set up the relevant Fokker–Planck equation for $P(x,t)$. What is (approximately) the escape rate for $kT = 0.1$ and mobility $\mu = 1$?

(b) Solve the PDE numerically starting from a narrow Gaussian distribution centered around $x = 0$ at $t = 0$, and compare with the result of (a). *Hint:* Choose a (fine) grid with spacing dx, and propagate the probability distribution with a small time step dt, choosing, for example, $dt = 0.1dx^2$ (can you see why this scaling is important?). A numerical solution with an accuracy within 20% is acceptable.

(c) What happens (qualitatively) to the probability distribution at long times? What would happen without the quadratic term in the potential?

5 Noise

Be not afeard; the isle is full of noises, sounds, and sweet airs, that give delight and
hurt not.

(William Shakespeare, The Tempest)

An important property of noise is its *power spectrum*, defined as

$$P(\omega) = \lim_{T \to \infty} \left\langle \frac{1}{T} \left| \int_0^T S(t) e^{-i\omega t} dt \right|^2 \right\rangle. \tag{5.1}$$

Here, $\langle\rangle$ denotes ensemble-averaging (averaging over multiple realizations of the disorder), though in most cases it is equal to averaging over different time windows (in
Chapter 7 we will encounter situations where these two are not equal to each other – a
phenomenon known as "ergodicity breaking"). It is not obvious a priori that the limit
exists – we shall shortly prove that. The dimensions of the power spectrum are $\frac{S^2}{\text{Hz}}$. It
is clear from the definition that $P(-\omega) = P(\omega)$, so there is no additional information
in the power spectrum at negative frequencies.

In atypical cases where the signal at hand is not a function of *time* but of, e.g., spatial
dimensions, the dimensions of the power spectrum will be, of course, different (with
Hz replaced by m^{-1}). For instance, one may consider the power spectrum in the context of natural images (Ruderman and Bialek 1994). Also note that in many practical
applications the signal analyzed is not continuous but discrete – it is straightforward
to generalize the results discussed in this chapter in the context of *continuous* signals
to discrete ones, see, for example, Porat (2008) and Problem 5.4.

Fig. 5.1 shows the (measured) estimated power spectrum of seismic fluctuations.

A basic relation between the power spectrum and the correlation function of the
noise is given by the Wiener–Khinchin Theorem (named after Norbert Wiener and
Aleksandr Khinchin), which will also show us why the limit exists.

We can write the power spectrum as

$$P(\omega) = \lim_{T \to \infty} \frac{1}{T} \int_0^T \int_0^T \left\langle S(t)S(t') \right\rangle e^{-i\omega t} e^{i\omega t'} dt dt', \tag{5.2}$$

where $C(t - t') = \left\langle S(t)S(t') \right\rangle$ is known as the autocorrelation of the signal.

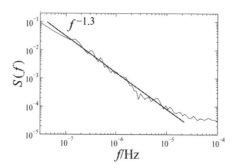

Figure 5.1 Example of a measured power spectrum estimation, in the context of seismology. Adapted Figure 4 with permission from Davidsen and Schuster (2002). Copyright 2002 by the American Physical Society.

We will shortly assume that the correlation function of the noise depends only on the difference $t - t'$. Such stochastic processes are known as wide-sense stationary.

Let us perform a change of variables, defining $t_1 = t - t'$ and $t_2 = t'$. The Jacobian of the transformation, J is

$$J = \begin{vmatrix} 1 & -1 \\ 0 & 1 \end{vmatrix} = 1. \tag{5.3}$$

This leads to

$$P(\omega) = \lim_{T \to \infty} \frac{1}{T} \int_0^T \int_{-t_2}^{T-t_2} C(t_1) e^{-i\omega t_1} dt_1 dt_2. \tag{5.4}$$

To proceed, let us denote the correlation time of the noise by τ, and assume $T \gg \tau$ (since we are interested in the limit $T \to \infty$). Whenever $t_2 \gg \tau$ the inner integral (over t_1) is well approximated by the Fourier transform of the correlation function (i.e., we can replace the limit $-t_2$ by $-\infty$ and $T - t_2$ by ∞ without significantly affecting the results). The fraction of values of t_2 in the range $[0, T]$ that violate this assumption scales as τ/T, and are thus negligible for $T \gg \tau$. We conclude that

$$P(\omega) = \int_{-\infty}^{\infty} C(t) e^{-i\omega t} dt. \tag{5.5}$$

Note that $C(-t) = C(t)$ since the autocorrelation function only depends on $\Delta t = |t - t'|$. It is therefore clear that Eq. (5.5) results in a real function, as it should.

What is the power spectrum good for? We have proved that the power spectrum is well defined, and it is one way to characterize a noisy process – but one may wonder what use it actually has. There are many answers to this question, on many levels, and we will name but a few. First, a mathematical characterization of the noise at a given frequency is crucial in order to know whether a sinusoidal signal buried in that noisy background can be detected (a common task in physics and engineering), and to determine for how long the measurement should be performed. Related to this, sensor performance is often characterized by providing the power spectrum of the sensor noise level, which determines the weakest signals that can be detected for a given measurement time. These aspects are explored in Problem 5.1. Furthermore, as we saw in the discussion of Chapter 3 on cell size control, information can be inferred from noise (see also Amir and Balaban (2018) for additional examples). Similarly, by quantifying the power spectrum associated with a given physical process we can infer valuable information regarding the underlying physical mechanisms. One beautiful example of this is shot noise – we will show that the power spectrum associated with a train of discrete pulses (e.g., conduction of electrons across a barrier such as a tunnel junction) is proportional to the product of the current and the area under the pulse (i.e., the *particle charge*). This implies that we can measure the charge of the particles by measuring the noise they produce – and indeed this approach has been successfully used to detect exotic phases of matter where the quasiparticle charge is different from the electron charge (Reznikov *et al.* 1999; Jehl *et al.* 2000). Similarly, the nature of the power spectrum itself can provide hints regarding the nature of the underlying mechanisms affecting the process – is the noise produced by stochastic jumps between two or more discrete states? (Such is the case for the "telegraph noise" we will study in detail.) Or a broad spectrum of rates? (Such is the case for the $1/f$ noise which is ubiquitous in nature.) Finally, the concept of power spectrum is useful for purely theoretical analysis. As a concrete example, we will obtain insights into the Ornstein–Uhlenbeck process by considering the power spectrum it generates.

5.1   Telegraph Noise: Power Spectrum Associated with a Two-Level-System

As an example (which has many applications), consider the noise arising from random jumps between two discrete values (say 0 and 1), as illustrated schematically in Fig. 5.2. The rate of going from 0 to 1 is γ_1, and the reverse rate is γ_2. This process is often referred to as "telegraph noise," since it involves transitions between two states.

To calculate the power spectrum, we will calculate the correlator

$$C(\Delta t) = \langle S(t)S(t') \rangle, \tag{5.6}$$

with $\Delta t = |t - t'|$. Without loss of generality, let us consider the case where $t > t'$. This is very similar to the calculations we did when dealing with "PageRank." We start with

$$\begin{pmatrix} \dot{p}_0 \\ \dot{p}_1 \end{pmatrix} = \begin{pmatrix} -\gamma_1 & \gamma_2 \\ \gamma_1 & -\gamma_2 \end{pmatrix} \begin{pmatrix} p_0 \\ p_1 \end{pmatrix} \tag{5.7}$$

Denoting $X \equiv p_0 + p_1$, it is clear that $\dot{X} = 0$. After a long time, an invariant distribution will be reached that is independent of the initial conditions. We have

$$0 = -\gamma_1 p_0 + \gamma_2 p_1 \rightarrow p_1 = p_0 \frac{\gamma_1}{\gamma_2}. \tag{5.8}$$

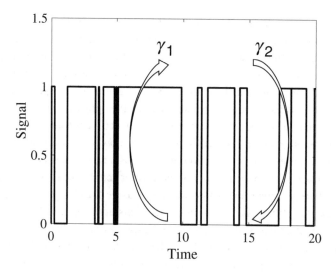

Figure 5.2 A simulation of a telegraph process. Here, $\gamma_1 = \gamma_2 = 1$.

Hence the stationary distribution is

$$(p_0, p_1) = \left(\frac{\gamma_2}{\gamma_1 + \gamma_2}, \frac{\gamma_1}{\gamma_1 + \gamma_2} \right). \tag{5.9}$$

From Eq. (5.7), we find that the characteristic polynomial is

$$(x + \gamma_1)(x + \gamma_2) - \gamma_1\gamma_2 = 0 \implies x^2 + (\gamma_1 + \gamma_2)x = 0. \tag{5.10}$$

Therefore, the two eigenvalues are

$$x_1 = 0 \text{ and } x_2 = -(\gamma_1 + \gamma_2). \tag{5.11}$$

The first corresponds to the stationary distribution, while the second will determine the dynamics (and hence the correlator and power spectrum).

If we start in state "1," the probability to be in it at time t is given by

$$p(t) = \frac{\gamma_1}{\gamma_1 + \gamma_2} + Ce^{-(\gamma_1+\gamma_2)t}, \tag{5.12}$$

where $C = \frac{\gamma_2}{\gamma_1+\gamma_2}$ to satisfy the initial conditions.

Since the correlator $\langle S(t)S(t')\rangle$ is only contributed by $S(t) = S(t') = 1$, we find that

$$\langle S(t)S(t')\rangle = \left(\frac{\gamma_1}{\gamma_1 + \gamma_2} \right) p(\Delta t) = Ae^{-(\gamma_1+\gamma_2)|t-t'|} + B, \tag{5.13}$$

with

$$A = \frac{\gamma_1\gamma_2}{(\gamma_1 + \gamma_2)^2}. \tag{5.14}$$

(Here, the term $\left(\frac{\gamma_1}{\gamma_1+\gamma_2} \right)$ arose from the steady-state probability of being in state "1.")

Now to get back the power spectrum we need to Fourier Transform. For finite frequencies the constant B will not contribute, and we find that

$$P(\omega) = \frac{\gamma_1 \gamma_2}{(\gamma_1 + \gamma_2)^2} \hat{f},$$

(5.15)

with

$$\hat{f} = \int_{-\infty}^{\infty} e^{-(\gamma_1 + \gamma_2)|t|} e^{-i\omega t} dt.$$

(5.16)

Evaluating the F.T. we find that

$$\hat{f} = -\frac{1}{-i\omega - \lambda} + C.C. = \frac{2\lambda}{\omega^2 + \lambda^2},$$

(5.17)

where $\lambda = \gamma_1 + \gamma_2$ and C.C. denotes the complex conjugate. Hence the power spectrum is a Lorentzian:

$$P(\omega) = A \frac{2\lambda}{\omega^2 + \lambda^2}.$$

(5.18)

An important limit is the one in which $\gamma_2 \gg \gamma_1$. In this case we find that for frequencies smaller than γ_2, we have

$$P(\omega) \approx 2A/\gamma_2 \approx 2\gamma_1/\gamma_2^2.$$

(5.19)

The lack of frequency dependence is known as *white noise*. Note that in this scenario the system will have short "pulses" in which it will be in the "1" state, of average duration $e = 1/\gamma_2$. This scenario is similar (but not identical) to the case of *shot noise*: In that case one has short pulses of constant duration, occurring at random times (e.g., imagine electrons flowing through a resistor). In our case the average "charge" is e, and the average current is $I = e\gamma_1$; therefore, the power spectrum is given by

$$P = 2eI,$$

(5.20)

which turns out to be the correct formula also in the case of shot noise – see Problem 5.2.

Let us repeat the analysis without relying on the Wiener–Khinchin relation, as we have done above, but by directly computing the Fourier transform of the signal and taking its squared magnitude. To simplify things, we will choose $\gamma_1 = \gamma_2 \equiv \hat{\lambda}$.

First, we note the relation of the Fourier transform of a signal versus that of its derivative. We have seen before that

$$\text{F.T.}[\dot{f}] = i\omega \text{F.T.}[f];$$

(5.21)

therefore,

$$P_{\dot{f}}(\omega) = \omega^2 P_f(\omega)$$

(5.22)

(where we used the notation P_g to denote the power spectrum of the signal g).

In the case where the signal jumps between 0 and 1, the derivative of the signal, $\dot{f}(t)$, will be composed of interlaced positive and negative δ-functions. When we Fourier transform $\dot{f}(t)$, we will obtain

$$P_{\dot{f}}(\omega) = \frac{1}{T}\left\langle \left|\int_0^T e^{-i\omega t} \sum_j \delta(t - t_j)(-1)^j dt\right|^2\right\rangle = \frac{1}{T}\left\langle \left|\sum_j e^{-i\omega t_j}(-1)^j\right|^2\right\rangle,$$

(5.23)

where t_j is the time of the j'th jump, and to avoid a cumbersome notation we left out the limiting process $\lim_{T\to\infty}$.

This can be rewritten as

$$P_{\dot{f}}(\omega) = \lim_{T\to\infty}\left\langle \frac{1}{T}\sum_{j,k} e^{-i\omega(t_j - t_k)}(-1)^{(j+k)}\right\rangle.$$

(5.24)

It would be convenient to separate the sum into two terms:

$$P_{\dot{f}}(\omega) = \left\langle \frac{1}{T}\sum_{j>k} e^{-i\omega(t_j - t_k)}(-1)^{(j+k)}\right\rangle + C.C. + \left\langle \frac{1}{T}\sum_{j=k} e^{-i\omega(t_j - t_k)}(-1)^{(j+k)}\right\rangle.$$

(5.25)

It is easy to see that the last term equals the average number of jumps in the interval, N, divided by T:

$$N/T \approx (T\hat{\lambda})/T = \hat{\lambda}.$$

(5.26)

Let us now evaluate:

$$I \equiv \left\langle \frac{1}{T}\sum_{j>k} e^{-i\omega(t_j - t_k)}(-1)^{(j+k)}\right\rangle.$$

(5.27)

Since we have a Markov process, the time difference $\tau_i \equiv t_{i+1} - t_i$ is distributed exponentially with a mean $1/\hat{\lambda}$. We can express the time differences between jumps (assuming $j > k$) as

$$t_j - t_k = \sum_{l=k}^{j-1} \tau_l.$$

(5.28)

Plugging this into the integral we find that

$$I \approx \sum_{j>k}\int_0^\infty\int_0^\infty ..\int_0^\infty \frac{1}{T}e^{-i\omega\left(\sum_{l=k}^{j-1}\tau_l\right)}(-1)^{(j+k)}P(\tau_1, \tau_2..\tau_N)d\tau_1..d\tau_N,$$

(5.29)

with $N = T\hat{\lambda}$ (note that in principle the number of jumps in the interval $[0, T]$ is a random variable, but for large T it is well approximated by N).

Since we have $P(\tau) = \hat{\lambda}e^{-\hat{\lambda}\tau}$, we obtain

$$I \approx (\hat{\lambda}^N) \sum_{j>k} \int_0^\infty \int_0^\infty \cdot\cdot \int_0^\infty \frac{1}{T} e^{-i\omega(\sum_{l=1}^{j-1} \tau_l)}(-1)^{(j+k)} e^{-\hat{\lambda}(\sum_{l=1}^{N} \tau_l)} d\tau_1..d\tau_N.$$

(5.30)

Consider the integral for a given j and k, with $j - k = l$. The integral over all τ_i which are not between j and k will be trivially one, and the remaining integral is

$$I_{jk} = \frac{1}{T}(\hat{\lambda}^l)(-1)^{(j+k)} \int_0^\infty \int_0^\infty \cdot\cdot \int_0^\infty e^{(-i\omega - \hat{\lambda})(\sum_{l=k}^{j-1} \tau_l)} d\tau_k..d\tau_{j-k}.$$ (5.31)

Evaluating the integral is straightforward and leads to

$$I_{jk} = \frac{(-1)^{(j+k)}}{T} \left(\frac{\hat{\lambda}}{\hat{\lambda} + i\omega} \right)^{j-k},$$

(5.32)

hence

$$I = \frac{1}{T} \sum_{j>k}(-1)^{(j+k)} \left(\frac{\hat{\lambda}}{\hat{\lambda} + i\omega} \right)^{j-k}.$$

(5.33)

Consider all of the (j, k) pairs with a separation $l > 0$. They give the same contribution, and their number is $N - l$, hence

$$I = \frac{1}{T} \sum_{l=1}^{N}(N - l)(-1)^l \left(\frac{\hat{\lambda}}{\hat{\lambda} + i\omega} \right)^l.$$

(5.34)

This is a geometric series, and will be dominated by the first few terms (where "few" is determined by the ratio of consecutive terms). Since we are interested in the limit of large N, for the dominant terms in the series we can replace $N - l$ with N, hence

$$I \approx -\frac{N}{T} \frac{\frac{\hat{\lambda}}{\hat{\lambda}+i\omega}}{1 + \frac{\hat{\lambda}}{\hat{\lambda}+i\omega}} = -\frac{\hat{\lambda}^2}{2\hat{\lambda} + i\omega}.$$

(5.35)

$$P_{\dot{f}}(\omega) = \frac{-\hat{\lambda}^2}{2\hat{\lambda} + i\omega} + C.C. + \hat{\lambda} = \frac{-4\hat{\lambda}^3}{4\hat{\lambda}^2 + \omega^2} + \hat{\lambda} = \frac{\omega^2 \hat{\lambda}}{4\hat{\lambda}^2 + \omega^2}.$$

(5.36)

Finally, we divide by ω^2 to find the spectrum of $p_f(\omega)$:

$$P_f(\omega) = \frac{\hat{\lambda}}{4\hat{\lambda}^2 + \omega^2},$$

(5.37)

which recovers Eq. (5.18), since $2\hat{\lambda} = \lambda$.

5.2 From Telegraph Noise to 1/f Noise via the Superposition of Many Two-Level-Systems

The telegraph noise arose from a process that jumped between two states, with two rates describing the transitions in either direction. This would come, for example, from two wells separated by a barrier, with transitions driven by thermal fluctuations leading to a rate approximately proportional to the Arrhenius form $e^{-U/kT}$ (as we discussed at length in Chapter 4). We just saw that such random transitions lead to a Lorentzian form of the power spectrum.

Consider now an ensemble of many such "two-level-systems," and assume a *distribution* of energy barriers. If that distribution is smooth on the scale of kT in the regimes of barriers associated with the experimentally observed timescales, we can approximate it by a constant. For this reason let's consider

$$P(U) \approx C \tag{5.38}$$

in the range $E_{min} < U < E_{max}$. This model has been used, for example, to describe the noise in the conductance of semiconductors due to impurities hopping between two states (Dutta and Horn 1981).

Let us first find the distribution of rates due to the distribution of energy barriers. This is similar to the calculations we did when discussing Benford's law. Since $\lambda \propto e^{-U/kT}$, we have

$$P(\lambda) = P(U)/|\frac{d\lambda}{dU}| = CkT/\lambda. \tag{5.39}$$

Note that the support of λ is the interval $[e^{-U_{max}/T}, e^{-U_{min}/T}]$.

A quick note on notation. Note that we have made a simple change of variables here and calculated the probability distribution of the new variable – similar to what we did when discussing Benford's law in Chapter 1. We have used the same notation P to describe the probability distribution, as is customary in the physics (but not mathematics) literature. This emphasizes that both random variables (which have distinct probability distributions) originate from the same stochastic process.

Next, we note that the sum of two independent random signals, $S = S_1 + S_2$, will have a power spectrum equal to the sum of the two power-spectra. This can be seen by considering the correlation function:

$$\langle S(t)S(0) \rangle = \langle S_1(t)S_1(0) \rangle + \langle S_2(t)S_2(0) \rangle + \langle S_1(t)S_2(0) \rangle + \langle S_2(t)S_1(0) \rangle. \tag{5.40}$$

Since the processes are independent, the cross-terms cannot depend explicitly on t, and hence will not contribute to the power spectrum. This leads to the power spectrum:

$$\int P(\lambda) \left(\frac{\lambda}{\lambda^2 + \omega^2} \right) d\lambda \propto \int_{\lambda_{min}}^{\lambda_{max}} \frac{T}{\lambda} \frac{\lambda}{\lambda^2 + \omega^2} d\lambda = T \arctan \left(\frac{\lambda}{\omega} \right) \frac{1}{\omega} \Big|_{\lambda_{min}}^{\lambda_{max}}. \tag{5.41}$$

Thus, for the range of frequencies $\lambda_{\min} \ll \omega \ll \lambda_{\max}$, we find that

$$P(\omega) \propto T/\omega. \tag{5.42}$$

Note that for $\omega \to 0$, we get saturation at a finite value, and for $\omega \to \infty$, we get a decay proportional to $\frac{1}{\omega^2}$ – hence the $1/f$ noise regime is only valid for intermediate asymptotics. Nevertheless, for a huge number of systems it is followed over a broad range of frequencies.

This mechanism can also be generalized beyond two-level-systems, to Markov processes corresponding to matrices with the appropriate distribution of eigenvalues (Amir, Oreg, and Imry 2009). For further discussion of other physical mechanisms leading to $1/f$, see also (Amir, Oreg, and Imry 2012) and references therein.

5.3 Power Spectrum of a Signal Generated by a Langevin Equation

Consider the following Langevin equation:

$$m\dot{v} + \eta v = f \tag{5.43}$$

(this is known as the Ornstein–Uhlenbeck process, and studied in Problem 3.3 of Chapter 3).

Let us find the power spectrum of $v(t)$, which will also tell us how the correlations decay in time. We have seen that

$$\langle f(t)f(t')\rangle = \Gamma\delta(t - t'). \tag{5.44}$$

This is known as *white noise*, since the Fourier transform of the this autocorrelation function is constant in frequency – akin to the color white, which is composed (uniformly) of all frequencies. However, the power spectrum of $v(t)$ will *not* be white, as we shall now see.

Taking the squared magnitude of the Fourier transform of the RHS of Eq. (5.43), ensemble-averaging and taking the limit $T \to \infty$ leads to its white-noise power spectrum:

$$\lim_{T \to \infty} \left\langle \frac{1}{T}\hat{f}(\omega)\hat{f}(-\omega) \right\rangle = \Gamma. \tag{5.45}$$

While following the same procedure for the LHS of Eq. (5.43) gives

$$\lim_{T \to \infty} \left\langle \frac{1}{T}[-i\omega m\hat{v}(\omega) + \eta\hat{v}][i\omega m\hat{v}(\omega) + \eta\hat{v}] \right\rangle. \tag{5.46}$$

(Note that in deriving this we utilized the fact that v is confined by the damping and does not diverge, so that we may neglect the boundary terms in the integration.)

Equating Eqs. 5.46 and 5.45 implies that

$$[\omega^2 m^2 + \eta^2] \lim_{T \to \infty} \left\langle \frac{1}{T}\hat{v}(\omega)\hat{v}(-\omega) \right\rangle = \Gamma. \tag{5.47}$$

Hence the velocity power spectrum is

$$P_v(\omega) = \frac{\Gamma}{\omega^2 m^2 + \eta^2}. \tag{5.48}$$

A *Lorentzian* power spectrum – implying that the correlation function of v decays exponentially. Hence we have

$$C(t) = \langle v(t)v(0) \rangle = Ae^{-\lambda t}, \tag{5.49}$$

and for consistency with Eq. (5.48) we have

$$\frac{A}{\lambda + i\omega} + C.C. = \frac{2A\lambda}{\lambda^2 + \omega^2} = \frac{\Gamma}{\omega^2 m^2 + \eta^2}, \tag{5.50}$$

leading to

$$A = \frac{\Gamma/m^2}{2\eta/m} = \frac{\Gamma}{2\eta m}; \lambda = \eta/m. \tag{5.51}$$

Note that since this is a physical description of a particle moving in a medium due to thermal fluctuations, we know that v should follow the Boltzmann distribution, i.e.,

$$p(v) \propto e^{-\frac{mv^2}{2kT}}. \tag{5.52}$$

This implies that

$$C(0) = \langle v^2 \rangle = \frac{kT}{m} = A. \tag{5.53}$$

For consistency, this leads to the relation:

$$\Gamma = 2\eta kT. \tag{5.54}$$

This is another example of a fluctuation–dissipation relation connecting the magnitude of the stochastic element (Γ) with the dissipation (η). In fact, we have seen before that Γ is related to the diffusion constant, via the relation $D = \Gamma/2$ (Eq. (4.13)), using which Eq. (5.54) is precisely the fluctuation–dissipation relation of Eq. (4.21) of Chapter 4.

5.4   Parseval's Theorem: Relating Energy in the Time and Frequency Domain

Consider the energy associated with the signal, defined as

$$E \equiv \int_{-\infty}^{\infty} f(t)^2 dt. \tag{5.55}$$

Representing $f(t)$ in terms of its Fourier transform, we can write this as

$$E = \frac{1}{(2\pi)^2} \int \left| \int \hat{f}(\omega)e^{i\omega t} d\omega \right|^2 dt \tag{5.56}$$

(where we omitted the limits of integration – which will always be $-\infty$ to ∞ in what follows).

This equals

$$\frac{1}{(2\pi)^2} \int \int \int \hat{f}(\omega)e^{i\omega t} \hat{f}(\omega')^* e^{-i\omega' t} dt\, d\omega\, d\omega'. \tag{5.57}$$

Noting that

$$\int \delta(t - t_0)e^{-i\omega t} dt = e^{-i\omega t_0}, \tag{5.58}$$

the inverse Fourier transform gives

$$\frac{1}{2\pi} \int e^{-i\omega t_0} e^{i\omega t} d\omega = \delta(t - t_0). \tag{5.59}$$

Performing the integration over time in Eq. (5.57), we find that

$$E = \frac{1}{2\pi} \int \hat{f}(\omega)d\omega\, \hat{f}^*(\omega)d\omega, \tag{5.60}$$

which is Parseval's theorem. In terms of $\omega = 2\pi\nu$, this can be written as

$$E = \int |f(\nu)|^2 d\nu. \tag{5.61}$$

In other words, we have shown that the energy in the time domain is mathematically related to that in the frequency domain. Going back to the power spectrum: If we consider the *power* (i.e., energy per unit time) associated with the signal, then we need to integrate $|f(t)|^2$ over a time window $[0, T]$ and then divide by T. Using Parseval's theorem, we see that the resulting power is proportional to the integral of the power spectrum: thus justifying its name!

Also note that in many cases the square of the signal is associated with a *physical* energy. Since the energy cannot be infinite in a physical system, one consequence of Parseval's theorem is that $1/f$ noise has to have a lower and upper cutoff – otherwise the energy in the frequency domain will diverge! This is indeed consistent with the model proposed earlier for $1/f$ noise, which indeed manifested these cutoffs and gave them a physical interpretation.

Working with Dirac's δ-function For non-physicists, working with entities such as the Fourier transform of $e^{i\omega t}$ may seem strange at first. Clearly, the integral defining it does not converge – so what does this mean? One way that may be helpful to think about it is to regularize the expression by multiplying it by $e^{-\epsilon|t|}$, with ϵ positive. Now the F.T. is well defined – let us denote it by $g(\omega)$ – and can readily be evaluated to give $\frac{2\epsilon}{\epsilon^2 + \omega^2}$, a Lorentzian. The smaller ϵ is, the narrower $g(\omega)$ is, yet the integral over ω is constant and equal to 2π. Thus, loosely speaking, as $\epsilon \to 0$ the F.T. becomes a δ-function (this can be made rigorous using distribution theory). This result could have been anticipated, as we did above, by noting that the inverse F.T. of a δ-function gives $\frac{1}{2\pi}e^{i\omega t}$.

5.5   Summary

This chapter introduced the concept of a *power spectrum*, which helps us characterize the properties of a noise – allowing us, for instance, to define the characteristics of a measurement device, or determine how long we should measure a signal to recover it from a noisy background. We saw that the Wiener–Khinchin theorem relates the original definition (in terms of the squared magnitude of the Fourier transform of the signal) to the Fourier transform of the noise autocorrelation function. We illustrated the calculations of a power spectrum on several pertinent examples: telegraph noise (and its corresponding Lorentzian power spectrum), $1/f$ noise (arising in our model from superimposing independent telegraph noise processes with a broad distribution of rates), as well as shot noise and white noise. We saw how the power spectrum can also aid theoretical calculations by analyzing the example of the Ornstein–Uhlenbeck process. Finally, we discussed Parseval's theorem relating the energy of a signal (or noise) in the time and spectral domain, giving us further intuition regarding the interpretation of the power spectrum.

For further reading There are many textbooks dealing with signal processing in far more detail than the brief introduction given here, especially in the context of digital signal processing (in contrast to the continuous description given here, in practice it is useful to develop the techniques for a discrete signal). See for example Porat (2008).

5.6   Exercises

5.1   Signal to Noise

(a)   An experimentalist tries to measure a perfectly sinusoidal voltage dependence on time, buried in a noisy signal. The noise is white and characterized by a power spectrum of N volts^2/Hz, while the signal has amplitude of A volts. How long (approximately) should she make the measurement?

(b)*   Realistic signals are not perfect. Assume now that the signal "dephases" (i.e., loses its phase coherence) over a characteristic time τ. For instance, you may assume that the signal is $A \sin(\omega t + \xi)$, where the phase ξ performs Brownian motion. Can the experimentalist succeed in measuring the signal in this case using the power spectrum? If so, how long a measurement time does she need?

5.2   Shot Noise

The discreteness of charge carriers implies that even for a constant current the random timing of the electron flow will give rise to noise which will depend on the magnitude of the current. Consider a system where the signal (current) occurs in δ-function pulses each with amplitude e the time of which is described by a Poisson process. What is the power spectrum of the noise?

5.3 Power Spectrum

Find an appropriate audio signal, and calculate numerically the following quantities:

(a) The power spectrum, by averaging over several windows.
(b) The correlation function, with cyclic boundary conditions.

5.4 Brown Noise and Beyond

(a) Let us define a 1D random walker by $x_{t+\Delta t} = (1 - \epsilon \Delta t)x_t + \sigma \xi_t$, where $\sigma > 0$ is a constant, ξ_t are independent random variables drawn from a normal distribution, with zero mean and unit variance, and $\epsilon > 0$ is a small, positive number. Find the power spectrum of x. You are free to consider time to be discrete or take the continuous-time limit $\Delta t \to 0$. If considering the continuous-time limit, identify how σ should scale with Δt.

(b) If $x_{t+\Delta t} = x_t + \sigma \xi_t$, find the power spectrum of x by calculating the Fourier transform of the signal. This is called "Brown noise," since it is the noise associated with Brownian motion.

(c) Compute the autocorrelation function for (b). Do the conditions of the Wiener–Khinchin theorem apply?

(d) If you take $\epsilon \to 0$ in part (a), do you recover the result of part (b)? Why or why not?

Consider now the process in part (a) with $1 - \epsilon \Delta t$ replaced by some constant λ between -1 and 1: $x_{t+\Delta t} = \lambda x_t + \sigma \xi_t$. In this case, time must be treated as discrete as the continuous-time limit does not exist. Let us take $\Delta t = 1$ for simplicity. For the discrete-time case, we define the power spectrum as $p(\omega) = \lim_{N \to \infty} \frac{1}{N} \left\langle \left| \sum_{n=0}^{N-1} x_n e^{-i\omega n} \right|^2 \right\rangle$.

(e)* Derive and prove the discrete version of the Wiener–Khinchin theorem.
(f)* Calculate the autocorrelation function corresponding to the discrete signal defined above.
(g)* Calculate the corresponding power spectrum.

5.5 Hypothetical Model for $1/f$ Noise

Consider an ensemble of a large number of capacitors. There is a small constant rate at which each of them is excited to a constant voltage, after which its current decays exponentially with rate λ_i, starting from an initial value $I = 1$. The signal is proportional to the sum of the currents of all capacitors.

(a) What is its power spectrum if all capacitors have the same λ?
(b) What distribution of λ is needed to obtain $1/f$ noise?

5.6 Tracking Colloidal Particles

Modern experiments using confocal microscopy are able to track the motion of individual sub-micron scale particles as a function of time. In some cases, these particles, which interact with each other, oscillate harmonically with small amplitudes around a given configuration. An experimentalist would like to find the frequencies of the normal modes of the vibrations. How can this be achieved in a robust way?

Hint: Consider the equal–time correlation matrix, and the results of Appendix B regarding correlations of Gaussian variables.

To see how these results can actually be implemented, see: Ghosh, A., Chikkadi, V. K., Schall, P., Kurchan, J., and Bonn, D., 2010. Density of states of colloidal glasses. *Physical Review Letters*, 104(24), p.248305. Chen, K., Ellenbroek, W. G., Zhang, Z., Chen, D. T., Yunker, P. J., Henkes, S., Brito, C., Dauchot, O., Van Saarloos, W., Liu, A. J. and Yodh, A. G., 2010. Low-frequency vibrations of soft colloidal glasses. *Physical Review Letters*, 105(2), p.025501.

5.7 Generalizing Telegraph Noise

Consider a system with three states, A, B, and C, which makes transitions at random according to a Markov chain with rate γ_{AB} from A to B and likewise between (A, C), (B, C). These transitions are also reversible, with the reverse rate denoted γ_{BA} and so forth. The transition rates are $\gamma_{AB} = 1$, $\gamma_{AC} = 2$, $\gamma_{BC} = 2$, $\gamma_{BA} = 3$, $\gamma_{CA} = 1$, and $\gamma_{CB} = 1$, in some appropriate units of $(\text{time})^{-1}$. There are no transitions from a state to itself. We can write a Fokker–Planck equation for the time-evolution of $\vec{p} = (p_A, p_B, p_C)$, a vector of probabilities that the system is in the three states, as

$$\frac{d\vec{p}}{dt} = M\vec{p}.$$

(a) What is the 3×3 matrix M, in terms of the rates? What is the stationary solution of $\vec{p}$?

(b) Find the eigenvalue–eigenvector pairs for M. Which one corresponds to the stationary state from (a)? Suppose that $\vec{p}(t = 0) = (1, 0, 0)$. Write this initial condition in the basis of eigenvectors, and hence give a formula for the subsequent time evolution of state A, $p_A(t)$. What is the rate of decay to the stationary state?

(c) Let $R(t)$ be the resistance as a function of time, which switches between the values $R_A = 1$, $R_B = 2$, and $R_C = 3$, and let $\overline{R}$ be the time average of the resistance. Determine the correlation function

$$C(t - t') = \left\langle (R(t) - \overline{R})(R(t') - \overline{R}) \right\rangle$$

explicitly. What is the limiting behavior of $C(t - t')$ for small and large $t - t'$?

(d) Compute the power spectrum of the resistance using the autocorrelation function found in (c).

This problem was adapted by Felix Wong from Problem 10.5 in Sethna, J., 2006. *Statistical Mechanics: Entropy, Order Parameters, and Complexity.* Oxford University Press.

5.8 Onsager Relations*

Consider the following Langevin equation:

$$\frac{d\vec{v}}{dt} = A\vec{v} + \vec{f}, \tag{5.62}$$

where A is a $N \times N$ matrix and $\vec{f}$ is a noise term.

(a) Assume that the noise is white (in time) but is correlated such that

$$\langle f_i(t_1) f_j(t_2) \rangle = C_{ij} \delta(t_1 - t_2), \tag{5.63}$$

with C a non-diagonal matrix. Express the correlation function $\langle v_i(t) v_j(0) \rangle$ and the power spectrum of v_i in terms of C and A.

(b) Assume that the free energy of the system can be written in the form

$$F = \sum_{i,j} M_{ij} v_i v_j. \tag{5.64}$$

(The probability to find the system in a given configuration v_i is proportional to $e^{-F/T}$.) Find the equal time correlation matrix $S \equiv \langle v_i(t) v_j(t) \rangle$.

(c) Express C in terms of the matrices A and S.

(d) Assuming time-reversal symmetry, prove that the matrix AS is a symmetric matrix. This symmetry is called the Onsager symmetry principle.

6 Generalized Central Limit Theorem and Extreme Value Statistics

I know of scarcely anything so apt to impress the imagination as the wonderful form of cosmic order expressed by the law of frequency of error.

(Sir Francis Galton, 1889, referring to the central limit theorem)

Let us consider some distribution $p(x)$ from which we draw random variables $x_1, x_2, \ldots, x_N$. If the distribution has a finite mean $\langle x \rangle$ and a finite standard deviation σ, then we can define

$$\xi \equiv \frac{\sum_{i=1}^{N}(x_i - \langle x \rangle)}{\sigma \sqrt{N}}. \tag{6.1}$$

By the central limit theorem the distribution of the scaled variable ξ approaches a Gaussian with vanishing mean and standard deviation of 1 as $N \to \infty$. What happens when $p(x)$ does *not* have a finite variance? Or a finite mean?

Perhaps surprisingly, in this case the Generalized Central Limit Theorem (GCLT) tells us that the limiting distribution belongs to a particular family (Lévy stable distributions), of which the Gaussian distribution is a proud member albeit e pluribus unum. Moreover, the familiar $\sqrt{N}$ scaling of the above equation does not hold in general, and its substitute will generally sensitively depend on the form of the *tail* of the distribution.

The results are particularly intriguing in the case of heavy-tailed distributions where the *mean* diverges. In that case the sum of N variables will be dominated by rare events, regardless of how large N is! Fig. 6.1c shows one such example, where a running sum of variables drawn from a distribution whose tail falls off as $p(x) \sim 1/x^{3/2}$ was used. The code which generates Fig. 6.1 is remarkably simple, and discussed in Section 6.7.

The underlying reason for this peculiar result is that for distributions with a power-law tail $p(x) \propto 1/x^{1+\mu}$, with $\mu \leq 1$, the distributions of both the *sum* and *maximum* of the N variables scale in the same way with N, namely as $N^{1/\mu}$ – dramatically different from the $\sqrt{N}$ scaling we are used to from the CLT. (Note that we are excluding the regime $1 < \mu < 2$ here, since in that case if the *mean* $\langle x \rangle$ of the distribution is nonzero, the dominant term in the running sum will be $N\langle x \rangle$ rather than $N^{1/\mu}$.) The distribution of the *maximum* is known as the *Extreme Value Distribution* or EVD (since for large N it inherently deals with rare, atypical events among the N i.i.d. variables). Surprisingly, also for this quantity universal statements can be made, and when appropriately scaled

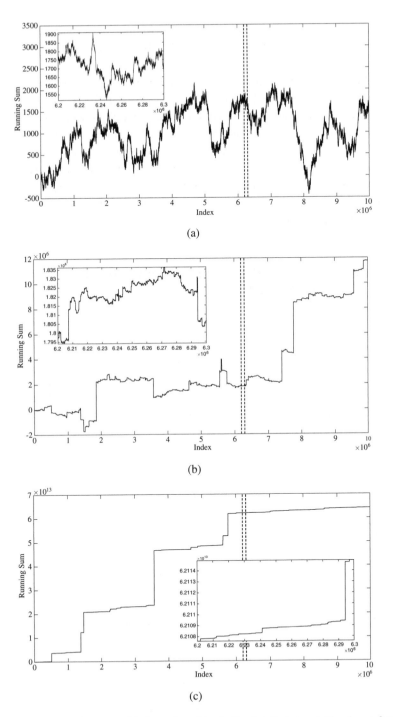

Figure 6.1 Running sum of independent, identically distributed variables drawn from three distributions: Gaussian (a), Cauchy (b, see Section 6.6 for its definition), and a distribution with positive support and a power-law tail $1/x^{3/2}$ (c). See Section 6.7 for details of the code. The insets illustrate the self-similar nature of the running sum, zooming into the small region of the original plot between the two vertical, dashed lines.

this random variable also converges to one of three universality classes – depending on the nature of the tails of the original distribution from which the i.i.d. variables are drawn.

In this chapter we will provide a straightforward derivation of these results. Some of the derivations presented follow closely that of Amir (2020), albeit with an extended (and more self-contained) description. Our derivations will not be mathematically rigorous – in fact, we will not even specify the precise conditions for the theorems to hold, or make precise statements about convergence. In this sense the derivation may be considered as "exact but not rigorous" (Feller (2008), for example, provides a rigorous treatment of many of the results derived in this chapter). Throughout, we will assume sufficiently smooth probability distributions, potentially with a power-law tail such that the variance or mean may diverge (known as a "fat" or "heavy" tail).

6.1 Probability Distribution of Sums: Introducing the Characteristic Function

An important quantity to define is the *characteristic function* of the distribution:

$$\varphi(\omega) = \int_{-\infty}^{\infty} f(t)e^{i\omega t} dt. \tag{6.2}$$

Clearly, $\varphi(0) = 1$, and we have $|\varphi(\omega)| < 1$ for $\omega \neq 0$ (can you see why?). Next, we shall derive a formula which will illustrate why the characteristic function will play a central role in this chapter.

Given two independent variables x, y with distributions $f(x)$ and $g(y)$, consider their sum $z = x + y$. The distribution of z is given by

$$p(z) = \int_{-\infty}^{\infty} f(x)g(z - x)dx, \tag{6.3}$$

i.e., it is the *convolution* of the two variables. Consider the characteristic function of the distribution:

$$\int_{-\infty}^{\infty} e^{i\omega z} p(z)dz = \int_{-\infty}^{\infty} \int_{-\infty}^{\infty} f(x)g(z - x)e^{i\omega z}dxdz. \tag{6.4}$$

Changing variables to x and $z - x \equiv y$, the Jacobian of the transformation is 1, and we find that

$$\int_{-\infty}^{\infty} e^{i\omega z} p(z)dz = \int_{-\infty}^{\infty} \int_{-\infty}^{\infty} f(x)g(y)e^{i\omega(x+y)}dxdy. \tag{6.5}$$

This is the product of the characteristic functions of x and y – which will shortly prove very useful.

This can be generalized to the case of the sum of n variables. Defining

$$X = x_1 + x_2 + \cdots + x_n, \tag{6.6}$$

we can write

$$\varphi_{sum}(\omega) = \int_{-\infty}^{\infty} \cdots \int_{-\infty}^{\infty} p(x_1)p(x_2)..p(x_n)dx_1\, dx_2 \,\ldots\, dx_n$$
$$\delta(x_1 + x_2 + \cdots + x_n - X)e^{i\omega X}dX, \tag{6.7}$$

which upon integrating over X leads to the product of the characteristic functions.

Finally, the same property also holds for the *Laplace* transform of positive variables, with essentially the same proof. Note that in that case, the upper cutoff of the convolution will be x rather than infinity.

Notation, notation ... Throughout this chapter we will (mostly) be using $p(x)$ to denote probability distributions and $\varphi(\omega)$ for the characteristic function. In cases where there is ambiguity regarding *which* variable the distribution is associated with, we will add a subscript, e.g., $p_n(x)$ refers to the probability distribution of the sum of n variables, and φ_{sum} is used above for the characteristic function of the sum. Towards the end of the chapter, cumulative distribution functions will be typically denoted by $C(x)$ or $G(x)$ (the latter notation will be used for the limiting distributions).

6.2   Approximating the Characteristic Function at Small Frequencies for Distributions with Finite Variance

Consider n i.i.d. variables drawn from a distribution $p(x)$, the characteristic function of which is $\varphi(\omega)$. The characteristic function of the sum is given by

$$\varphi_n(\omega) = \big(\varphi(\omega)\big)^n. \tag{6.8}$$

Let us assume that the distribution has a finite variance σ^2 (and hence finite mean). We shall see that the fact that the variance exists implies that the characteristic function attains a quadratic form near the origin.

Without loss of generality let us assume that the mean is 0 – otherwise we can always consider a new variable with the mean subtracted. We can Taylor expand $e^{i\omega x} \approx 1 + i\omega x - \frac{\omega^2}{2}x^2$ near the origin, and find that

$$\varphi(\omega) \approx 1 + a\omega + b\omega^2 + \cdots, \tag{6.9}$$

where a clearly vanishes and

$$b = -\frac{1}{2}\int_{-\infty}^{\infty} x^2 p(x)dx = -\frac{\sigma^2}{2}. \tag{6.10}$$

Therefore,

$$\varphi_n(\omega) \approx \left(1 - \frac{\sigma^2\omega^2}{2}\right)^n. \tag{6.11}$$

We can now write, following the logic of Eq. (2.53),

$$\varphi_n(\omega) \approx e^{n \log\left(1 - \frac{\sigma^2 \omega^2}{2}\right)} \approx e^{-n \frac{\sigma^2 \omega^2}{2}}. \tag{6.12}$$

In the next section we shall see the significance of the correction to the last expansion. Taking the inverse Fourier transform leads to

$$p_n(x) \propto e^{-\frac{x^2}{2\sigma^2 n}}, \tag{6.13}$$

i.e., the sum is approximated by a Gaussian with standard deviation $\sigma \sqrt{n}$, as we know from the CLT. You may worry about the validity of the approximations we made above: Indeed, in the next section we will see that the approximation we made is a good one only within a finite region around the peak of the Gaussian (called the central region), and we will determine the width of this region in certain cases.

Note that the finiteness of the variance – associated with the properties of the *tail* of $p(x)$ – was essential to establishing the behavior of $f(\omega)$ for *small* frequencies. This is an example of a Tauberian theorem, where large x behavior determines the small ω behavior, and we shall see that this kind of relation holds also for cases when the variance diverges.

Before going to these interesting scenarios, let us consider a particular example with finite variance, and find the regime of validity of the central limit theorem.

6.3   Central Region of CLT: Where the Gaussian Approximation Is Valid

Consider the distribution,

$$p(x) = e^{-x} \theta(x), \tag{6.14}$$

where $\theta(x)$ is the Heaviside step function (1 for positive arguments, 0 otherwise).

It is easy to calculate its characteristic function, finding that

$$\varphi(\omega) = \frac{1}{1 - i\omega}. \tag{6.15}$$

Hence

$$\varphi^n(\omega) = \left(\frac{1}{1 - i\omega}\right)^n. \tag{6.16}$$

Thus, the distribution of the sum, $p_n(x)$ is given by the inverse Fourier transform as

$$p_n(x) = \frac{1}{2\pi} \int_{-\infty}^{\infty} e^{-i\omega x} \left(\frac{1}{1 - i\omega}\right)^n d\omega. \tag{6.17}$$

We can take the inverse Fourier by evaluating a simple contour integration, with a single pole of order n at $z_0 = -i$ (see Appendix C for a brief description of higher order poles and their residues). For negative x, we can close the contour from above, hence the integral vanishes. For positive x, we can close the contour from below, and find that integral is contributed by the complex pole, giving a contribution of

$$I = 2\pi i \frac{r^{(n-1)}(z_0)}{n-1!}, \tag{6.18}$$

where the function $r(z) = (i^n)e^{-izx}$. Note that since the contour integration is done in the clockwise direction, we obtain that the integral of Eq. (6.17) equals to $-I$. Putting it all together we find the following distribution function for the sum of n variables drawn from the distribution $p(x)$:

$$p_n(x) = \theta(x)e^{-x}\frac{x^{n-1}}{n-1!} \propto \theta(x)e^{(n-1)\log(x)-x}. \tag{6.19}$$

Note that its average is clearly n (since the average of the original variable is 1). The maximum of this function occurs at $x = n - 1$.

A different interpretation The form of Eq. (6.19), as a function of the integer n, is that of the Poisson distribution – this is not a coincidence and we could have bypassed the calculation of the above contour integration by the following probabilistic argument (by Yipei Guo): given a Poisson process with rate r (with $r = 1$ for the example above), let us denote by n_t the number of events happening between time $t = 0$ and time t (a random variable of course). Denote by t_n the sum of the durations for the first n events – the distribution of which, $p_n(t_n)$, is precisely what we calculate above and given by Eq. (6.19). Noting that the event $t_n < t$ is the complementary of the event $n_t < n$ leads to

$$p_n(t_n) = \frac{d\text{Prob}[t_n < t]}{dt} = -\frac{d\text{Prob}[n_t < n]}{dt} = r\text{Prob}[n_t = n - 1], \tag{6.20}$$

and by definition $\text{Prob}[n_t = n - 1]$ is given by the Poisson distribution.

Let us expand the distribution around the maximum. We will define $g(x) \equiv (n-1)\log(x) - x$. Clearly, the first derivative vanishes, and expanding the exponent to *third* order we find that

$$g(x) \approx (n-1)\log(n-1) - (n-1) + \frac{g''}{2}(\delta x)^2 + \frac{g'''}{6}(\delta x)^3, \tag{6.21}$$

with $\delta x = x - (n-1)$.

The first two terms will lead to a Gaussian behavior – as expected by the CLT. We have

$$g''(x) = -(n-1)/x^2, \tag{6.22}$$

hence $g''(n) \approx -1/n$, and the behavior near the maximum is approximated by

$$p_n(x) \propto e^{-\frac{(\delta x)^2}{2n}}, \tag{6.23}$$

a Gaussian with standard deviation n. This had to be the case since the variance of the original distribution is

$$\int_0^\infty e^{-x}x^2 dx - \left(\int_0^\infty e^{-x}x \, dx\right)^2 = 2 - 1 = 1. \tag{6.24}$$

The multiplicative correction C to the Gaussian will be a factor:

$$C \equiv e^{\frac{g'''}{6}(\delta x)^3}, \tag{6.25}$$

hence notable corrections will occur at sufficiently large δx such that

$$\delta x \sim (g''')^{-1/3}. \tag{6.26}$$

Since

$$g'''(x) = 2(n-1)/x^3 \implies g'''(n) \approx 2/n^2, \tag{6.27}$$

the regime of the CLT is of width

$$\delta x \sim n^{2/3}. \tag{6.28}$$

The essential thing is that this is parametrically larger than the width of the Gaussian ($\sqrt{n}$), hence when the deviations begin the distribution has already decayed dramatically from its peak value – we will be in the tails of the Gaussian.

Furthermore, if the function was symmetric around the maximum the g''' term would have vanished, and we would have needed to go to the fourth order derivative, which would scale as n^{-3}. This would lead to a scaling:

$$\delta x \sim n^{3/4}. \tag{6.29}$$

This is explored in Problems 6.1 and 6.2 and related to the so-called Large deviation formalism in Problem 6.3 (see also Schuss 2009; Touchette 2009). Hence, as expected, the width of the central region is much larger for symmetric functions. Note that these two scaling relations are generic – as long as the tail of the distribution decays sufficiently fast. In Problems 6.4 and 6.5 you will see that the width of the central region is much smaller in cases where the function has a power-law tail – even when it has a finite variance.

More general approaches to the central region Given the importance of the central limit theorem it is not surprising that more general and sophisticated methods have been developed for determining the central region. Feller (2008); Lam et al. (2011) and Hazut et al. (2015) provide a good overview (and some recent extensions) of these methods, including the so-called Edgeworth expansion and the Gram–Charlier expansion (see also https://ocw.mit.edu/courses/mathematics/18-366-random-walks-and-diffusion-fall-2006/lecturenotes/lec03.pdf.)

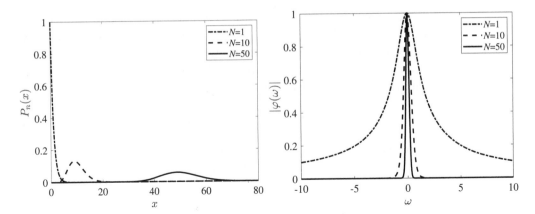

Figure 6.2 The sum of N random variables, each distributed exponentially for various values of N (left, see Eq. (6.19)), and the corresponding magnitude of the characteristic functions (right, see Eq. (6.16)).

Finally, we should emphasize that these scaling laws go beyond the central limit theorem, which only tells us that the scaled variable

$$\zeta_N = \frac{s_N - \langle x \rangle N}{\sigma \sqrt{N}} \tag{6.30}$$

converges to a normalized Gaussian distribution in the limit $N \to \infty$, which is equivalent to the statement that the central region grows faster than $\sqrt{N}$ – but does not tell us how much faster.

We can also use the above example to illustrate a general principle we will utilize repeatedly in this chapter: the reciprocity of real space and Fourier space, related to the effectiveness of the Tauberian theorems we will use (an example of which was mentioned earlier). This is illustrated in Fig. 6.2. As we increase N, the function becomes broader in real space, but narrower in Fourier space. Since in Fourier space the characteristic function is simply raised to the power of N, a large N makes it narrower and narrower, still equal to 1 at $\omega = 0$ but decreasing more and more rapidly in magnitude as a function of ω as we increase N. For this reason taking large N essentially "zooms into" the small ω region, which implies that only the *tails* of the function in real space matter (since they determine the characteristic function at small ω). This notion is at the heart of the universal behavior we will explore in this chapter.

6.4   Sum of a Large Number of Positive Random Variables: Universal Description in Laplace Space

We will first consider the case of n i.i.d. variables drawn from a distribution $p(x)$ with positive support and whose tail falls off as a power-law, i.e., from a point x^* onwards $p(x) = A/x^{1+\mu}$. The derivation follows closely that of Bardou *et al.* (2002). Clearly, $\mu > 0$, in order for the distribution to be normalizable. If $\mu > 2$, the CLT holds, and the sum converges to a Gaussian. In this section, we will focus on the case $0 < \mu < 1$. In this case the distribution has neither a finite mean nor variance. We would like to consider the Laplace transform of the distribution, $L(s)$, for small values of s. However, if we were to naively expand e^{-sx} for small s, already the first-order term in the expansion would lead to a divergence:

$$\int_0^\infty p(x)[1 - sx]dx = \infty. \tag{6.31}$$

The trick will be to write the Laplace transform as

$$L(s) = \int_0^\infty p(x)[1 + e^{-sx} - 1]dx = 1 + \int_0^\infty p(x)[e^{-sx} - 1]dx. \tag{6.32}$$

We will now separate the integral into two regions: one from 0 to x^*, and the second from x^* to ∞, namely

$$I_1 \equiv \int_0^{x^*} p(x)[e^{-sx} - 1]dx \tag{6.33}$$

and

$$I_2 \equiv \int_{x^*}^\infty p(x)[e^{-sx} - 1]dx. \tag{6.34}$$

We will show that I_1 can be at most linear in s, while I_2 scales as s^μ for $\mu < 1$ and as $-s \log(s)$ for $\mu = 1$, and is therefore dominant. Using the simple inequality $|e^{-sx} - 1| \le sx$, we find that

$$|I_1| \le \int_0^{x^*} p(x)sxdx \le sx^*. \tag{6.35}$$

The second term will be

$$I_2 = \int_{x^*}^\infty \frac{A}{x^{1+\mu}}[e^{-sx} - 1]dx. \tag{6.36}$$

Performing the integration by parts, we find that

$$I_2 = -\frac{A}{\mu}x^{-\mu}[e^{-sx} - 1]\Big|_{x^*}^\infty - \int_{x^*}^\infty \frac{A}{\mu}x^{-\mu}se^{-sx}dx. \tag{6.37}$$

The first term can be bounded in magnitude by

$$E = \frac{A}{\mu}(x^*)^{-\mu}sx^*, \tag{6.38}$$

and hence is at most linear in s. The second term can be evaluated by a simple change of variables $sx \equiv m$, leading to

$$\int_{sx^*}^\infty \frac{A}{\mu}m^{-\mu}s^\mu se^{-m}dm/s = \frac{As^\mu}{\mu}\int_{sx^*}^\infty m^{-\mu}e^{-m}dm. \tag{6.39}$$

In the limit $sx^* \to 0$, this integral becomes the gamma function:

$$\Gamma(a) = \int_0^\infty x^{a-1}e^{-x}dx. \tag{6.40}$$

For $\mu < 1$, the integral will converge also when we set the lower limit to be 0, and its value will be hardly changed, hence we can approximate it by $\Gamma(1 - \mu)$.

Putting it all together we find that for $\mu < 1$ the Laplace transform is approximately

$$L(s) = 1 + I_1 + I_2 \approx 1 - \frac{A\Gamma(1 - \mu)}{\mu} s^\mu + O(s). \tag{6.41}$$

This is another example of a Tauberian theorem: the behavior of the tail of $p(x)$ determined the behavior of $L(s)$ for small values of s.

What would happen if we rescale the random variable by a positive constant a? The new variable $y = ax$ has a probability distribution $\frac{1}{a} p(y/a)$; therefore, its Laplace transform is

$$\int_0^\infty \frac{1}{a} p(y/a) e^{-sy} dy = \int_0^\infty p(x) e^{-sax} dx = L(sa) \tag{6.42}$$

(i.e., the broader a function is in real space the narrower it is in Fourier/Laplace space). Defining $C = \frac{A\Gamma(1-\mu)}{\mu}$, Eq. (6.41) implies that for the rescaled variable,

$$L(s) \approx 1 - Ca^\mu s^\mu + O(s). \tag{6.43}$$

Consider now a sum of N i.i.d. variables drawn for $p(x)$, and consider rescaling the sum by a negative power of N, $1/N^\alpha$, where $\alpha > 0$ is yet to be determined.

We have

$$L_N(s) \approx \left[1 - \frac{Cs^\mu}{N^{\alpha\mu}} \right]^N. \tag{6.44}$$

It is now evident that if we choose $\alpha = 1/\mu$, the *rescaled* sum would converge to the limit:

$$L_\infty(s) = e^{-Cs^\mu}. \tag{6.45}$$

Taking the inverse Laplace transform will determine the probability distribution of the rescaled sum. In other words, if we take the sum of the n variables and scale it by a factor $n^{1/\mu}$, it will converge to a particular distribution as $n \to \infty$. This is distinctly different from the $\sqrt{n}$ scaling associated with the CLT – here the scaling is superlinear in n! You will explore this intriguing fact numerically in Problem 6.6.

Limiting distributions What we just showed is that upon appropriate scaling, the rescaled sum converges to a limiting distribution for large N. This is analogous to the central limit theorem, albeit in the latter case a shift by the mean is required (which we didn't require above – the mean is ill-defined!) and the scaling factor scales as $\sqrt{N}$ rather the power–law behavior we found. Nevertheless, both of these cases exemplify the concept of a limiting distribution – upon appropriate shifts and scaling, the sum converges to an N-independent distribution.

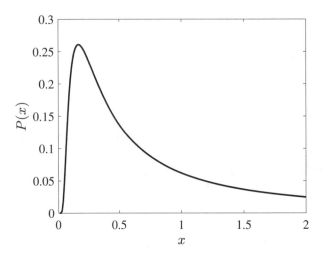

Figure 6.3 Lévy stable distribution for a strictly positive variable with $\mu = 0.5$.

As a concrete example, consider the case $\mu = 1/2$. We can express the inverse Laplace analytically as

$$P(x) = \frac{1}{\sqrt{4\pi}} \frac{e^{-\frac{1}{4x}}}{x^{3/2}}. \tag{6.46}$$

This particular function is known as the Lévy distribution (named after Paul Lévy). Note that it has the same power-law tail as the original distribution. This is shown in Fig. 6.3.

In Section 6.8 we will use a different approach to find the form of the limiting distribution for $\mu \geq 1$, in Fourier space rather than Laplace space. This is in fact essential, since in this case the limiting distribution will have support also for negative values.

6.5   Application to Slow Relaxations: Stretched Exponentials

Gaussians are very common – since they may arise from a sum of a large number of independent random variables drawn from *any* distribution with a finite variance. The other stable distributions are also quite common and useful to many applications, since they only depend on the tail of the distribution, and power-law distributions are quite common in nature.

Early on in our studies of science we are exposed to examples of *exponential* relaxations – a capacitor discharging through a resistor, or a parachuter reaching terminal velocity. Yet in numerous applications the relaxations are non-exponential, and have a very slow tail.

It is possible to obtain such slow relaxations by taking the superposition of many decaying exponentials. In Chapter 5, when discussing $1/f$ noise in semiconductors, we considered a model which presented a distribution of relaxation rates $p(\lambda) \propto 1/\lambda$

(with lower and upper cutoffs to make the distribution normalizable). This came about from thermal activation over barriers with an approximately uniform distribution, and led to a superposition of *Lorentzians* that gave $1/f$ noise over a broad range. Superimposing many exponential relaxations with rates drawn from that distribution gives us

$$S(t) \propto \int_{\lambda_{min}}^{\lambda_{max}} \frac{C}{\lambda} e^{-\lambda t} d\lambda. \qquad (6.47)$$

We have seen this integral before in Chapter 2, in the context of the *well function* – it can be written in terms of the exponential integral function, and diverges logarithmically at long times (but short compared to $1/\lambda_{min}$):

$$S(t) \approx -\gamma_E - \log(\lambda_{min} t). \qquad (6.48)$$

Logarithmic relaxations are extremely common in nature, as shown on two very different examples in Fig. 6.4. Note that we have also seen another route to obtain a distribution closely related to the $1/\lambda$ distribution when discussing Benford's law in Chapter 1: a multiplicative process resulting in a log-normal distribution. This mechanism would therefore also give approximately logarithmic relaxations, without relying on thermal activation (Amir, Oreg, and Imry 2012).

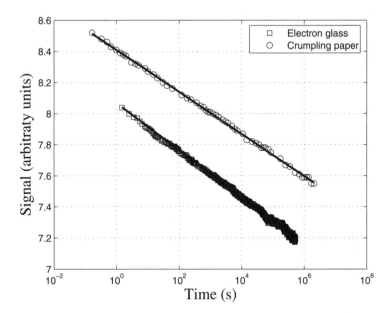

Figure 6.4 Experimental results showing a logarithmic relaxation in the electron glass indium oxide, where conductance is measured (Orlyanchik and Ovadyahu, 2004), and in a crumpled sheet, where the height is measured (Matan, Williams, Witten, and Nagel, 2002), after a sudden change in the experimental conditions. As seen in the graph, the logarithmic change in the physical observable can be measured from times of order of seconds or less to several days. Similar logarithmic relaxations, observed over many decades in time, occur in numerous physical systems, ranging from currents in superconductors to frictional systems. From Amir, Oreg, and Imry, *PNAS* 109, 6, 1850 (2012).

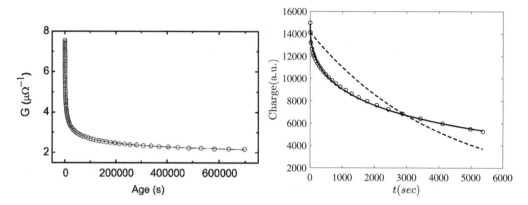

Figure 6.5 (left) Experimental results for relaxation of the conductance of a tunnel junction. Adapted Figure 1 with permission from Nesbitt and Hebard (2007). Copyright 2007 by the American Physical Society. The fit is to a stretched exponential function, $f(t) \propto e^{-(t/\tau)^\alpha}$. (right) The first observation of a stretched exponential in nature was in measurements of charge relaxations in a Leiden jar by R. Kohlrausch, reported in *Annals of Physics and Chemistry* (Poggendorff) 91, 179 (1854). The solid line shows a stretched exponential, while the dashed line shows exponential decay.

Another class of relaxations which have been shown to arise in nature already in the nineteenth century are *stretched exponentials*. In this case the measured quantity decays as

$$S(t) \propto e^{-Ct^\mu}, \tag{6.49}$$

with $\mu < 1$: see Fig. 6.5 for two examples.

How can we understand these relaxations? In analogy with Eq. (6.47), we would like to write them as the Laplace transform of a hopefully "natural" distribution of relaxation rates. Taking the inverse Laplace transform of Eq. (6.49) is not feasible analytically, but in fact, we already know which distributions lead to this form when their Laplace transform is taken: These are precisely the distributions discussed in section 6.4 associated with a positive variable (indeed, the relaxation rates must all be positive). Since they arise whenever a sum of variables drawn from a power-law tailed distribution is involved, they may help us rationalize the abundance of stretched exponential relaxations in nature. See also Montroll and Shlesinger (1983) and references therein for an extended discussion.

6.6   Example of a Stable Distribution: Cauchy Distribution

Stable distributions In the previous section, we say that the sum of (positive) variables drawn from a heavy-tailed distribution converges to a distribution whose shape is given by the inverse Laplace transform of the function e^{-s^μ}. Considering two variables drawn from this limiting distribution, the Laplace transform of the sum is clearly

e^{-2s^μ} – hence the resulting distribution is identical up to rescaling. This is an example of a Lévy stable distribution. In this case its support was strictly positive, but we have also seen another example where this is not the case: the Gaussian distribution. Next, we shall see another example of a symmetric Lévy-stable distribution, albeit with a heavy-tail, and in the next section we will find *all* Lévy stable distributions. Finally, it is worth commenting that every stable distribution is clearly its own limiting distribution. As may be anticipated, the converse is also true: Every limiting distribution must be stable.

Consider the following distribution, known as the Cauchy distribution (named after Augustin-Louis Cauchy):

$$P(x) = \frac{1}{\gamma\pi(1 + (\frac{x}{\gamma})^2)} \qquad (6.50)$$

(note that this has the same Lorentzian expression as telegraph noise). Its Fourier transform is

$$\varphi(\omega) = e^{-\gamma|\omega|}. \qquad (6.51)$$

Thus the characteristic function of a sum of N such variables is

$$\varphi_N(\omega) = e^{-N\gamma|\omega|}, \qquad (6.52)$$

and taking the inverse Fourier transform we find that the sum also follows a Cauchy distribution:

$$p_N(x) = \frac{1}{\gamma\pi N(1 + (\frac{x}{N\gamma})^2)}. \qquad (6.53)$$

Thus, the sum does not converge to a Gaussian, but rather retains its Lorentzian form. Moreover, it is interesting to note that the scaling form governing the width of the Lorentzian evolves with N in a different way than the Gaussian scenario: While in the latter the variance increases linearly with N hence the width increases as $\sqrt{N}$, here the scaling factor is linear in N. This remarkable property is in fact useful for certain computer science algorithms (Indyk 2000).

6.7 Self-Similarity of Running Sums

To generate Fig. 6.1, we generate a set of i.i.d. variables from a given distribution (Gaussian, Cauchy, and a heavy-tailed distribution with positive support whose tail falls off as $1/x^{3/2}$). For each long sequence of random variables, the running sum is plotted. For the Gaussian case (or any case where the variance is finite), the result is the familiar process of diffusion: The variance increases linearly with "time" (i.e., the index of the running sum). For Fig. 6.1a the mean vanishes, hence the running sum follows this random walk behavior. If we were to repeat this simulation many times, the result of the running sum at time N, a random variable of course, is such that when scaled by $1/\sqrt{N}$ it would follow a normal distribution with variance 1, as noted in Eq. (6.1). Another important property is that "zooming" into the running sum (see the figure inset) looks identical to the original figure – as long as we don't zoom in too far as to reveal the granularity of the data. Fig. 6.1b shows the same analysis for the Cauchy distribution of Eq. (6.50). As we have seen, now the scaling is linear in N. Nevertheless, zooming into the data still retains its Cauchy statistics. The mathematical procedure we will shortly follow to find *all* Lévy stable distributions will rely on this self-similarity. Indeed, assume that a sum of variables from some distribution converges – upon appropriate linear scaling – to some Lévy stable distribution. Zooming further "out" corresponds to generating sums of Lévy stable variables, hence it retains its statistics. A dramatic manifestation of this is shown in Fig. 6.1c. The initial distribution is *not* a Lévy stable, but happens to have a very fat tail, possessing infinite mean and variance. The statistics of the running sum converges to a Lévy stable distribution – in this case fortuitously expressible in closed form, corresponding to the Lévy distribution of Eq. (6.46). Importantly, zooming into the running sum still retains its statistics, which in this case happens to manifest large jumps associated with the phenomenon of Lévy flights, which will be elucidated by our later analysis. Due to the self-similar nature, zooming into what seems to be flat regions in the graph shows that their statistical structure is the same, and they also exhibit these massive jumps.

In order to generate the data shown in Fig. 6.1(a), the following MATLAB code is used:

```
tmp=randn(N,1);
x=cumsum(tmp);
```

For Fig. 6.1(b), to generate a running sum of variables drawn from the Cauchy distribution, the first line is replaced with

```
tmp = tan(pi*(rand(N,1)-1/2));
```

(Can you see why this works?)

Finally, for Fig. 6.1(c) the same line is replaced with

```
t=rand(N,1);
b=1/mu;
tmp=t.^(-b);
```

where we used $\mu = 1/2$ for the figure.

6.8 Generalized CLT via an RG-Inspired Approach

We shall now use a different method to generalize the result further, to distributions with arbitrary support. We will look for distributions which are *stable*: This means that if we add two (or more) variables drawn from this distribution, the distribution of the sum will retain the same shape – i.e., it will be identical up to a potential shift and scaling, by some yet undetermined factors. If the sum of a large number of variables drawn from *any* distribution converges to a distribution with a well-defined shape, it must be such a stable distribution. The family of such distributions is known as *Lévy stable*.

Clearly, a Gaussian distribution obeys this property. We also saw that the inverse Laplace transform of e^{-s^μ} with $0 < \mu < 1$ obeys it. We shall now use an RG (renormalization group) approach to find the general form of such distributions, which will turn out to have a simple representation in Fourier rather than real space – because the characteristic function is the natural object to deal with here. We also note that related renormalization-group-inspired ideas have been utilized in the context of the "conventional" CLT (see, for example, exercise 12.11 in Sethna 2006). See also Amir (2020) for a discussion of the distinction of the RG-inspired idea utilized in the following and other RG-approaches to the CLT (Jona-Lasinio 2001; Calvo 2010).

The essence of the approach relies on the fact that if we sum a large number of variables, and the sum converges to a stable distribution, then by definition taking a sum involving, say, twice the number of variables, will *also* converge to the same distribution – up to a potential shift and rescaling. This is illustrated visually in Fig. 6.1 by plotting the running sum of independently and identically distributed variables.

Defining the partial sums by s_n, the general (linear) scaling one may consider is

$$\xi_n = \frac{s_n - b_n}{a_n}. \tag{6.54}$$

Here, a_n determines the width of the distribution, and b_n is a shift. If the distribution has a finite mean it seems plausible that we should center it by choosing $b_n = \langle x \rangle n$. We will show that this is indeed the case, and that if its mean is infinite we can set $b_n = 0$. Note that we have seen this previously for the case of positive variables.

The scaling we are seeking is of the form of Eq. (6.54), and our hope is that if the distribution of ξ_n is $p_{\xi_n}(x)$, then

$$\lim_{n \to \infty} p_{\xi_n}(x) = p(x) \tag{6.55}$$

exists, i.e., the scaled sum converges to some distribution $p(x)$, which is not necessarily Gaussian (or symmetric).

Let us denote the characteristic function of the scaled variable by $\varphi(\omega)$ (assumed to be approximately independent of n for large n). Consider the variable $y_n = \xi_n a_n$. We have

$$p_{y_n}(y_n) = \frac{1}{a_n} p(y_n / a_n), \tag{6.56}$$

with p the limiting distribution of Eq. (6.55) (and the factor $\frac{1}{a_n}$ arising from the Jacobian of the transformation). Therefore, by the same logic of Eq. (6.42), the characteristic function of the variable y_n is

$$\varphi_{y_n}(\omega) = \varphi(a_n \omega). \tag{6.57}$$

Note that the $\frac{1}{a_n}$ prefactor canceled, as it should: since the characteristic function must be 1 at $\omega = 0$. Consider next the distribution of the sum s_n. We have $s_n = y_n + b_n$, and

$$p_{s_n}(s_n) = p_{y_n}(s_n - b_n). \tag{6.58}$$

Shifting a distribution by b_n implies multiplying the characteristic function by $e^{i\omega b_n}$. Therefore, the characteristic function of the sum is

$$\varphi_{s_n}(\omega) = e^{ib_n\omega}\varphi_{y_n}(\omega) = e^{ib_n\omega}\varphi(a_n\omega). \tag{6.59}$$

This form will be the basis for the rest of the derivation, where we emphasize that the characteristic function φ is n-independent.

Consider $N = n \cdot m$, where n, m are two large numbers. The important insight is to realize that one may compute s_N in two ways: as the sum of N of the original variables, or as the sum of m variables, each one being the sum of n of the original variables. The characteristic function of the sum of n variables drawn from the original distribution is given by Eq. (6.59).

If we take a sum of m variables drawn from *that* distribution (i.e., the one corresponding to the sums of n's), then its characteristic function will be on the one hand

$$\varphi_{s_N}(\omega) = e^{imb_n\omega}(\varphi(a_n\omega))^m, \tag{6.60}$$

and on the other hand it is the distribution of $n \cdot m = N$ variables drawn from the original distribution, and hence does not depend on n or m separately but only on their product N. Therefore, assuming that n is sufficiently large such that we may treat it as a continuous variable, we have

$$\frac{\partial}{\partial n}e^{i\frac{N}{n}b_n\omega+\frac{N}{n}\log[\varphi(a_n\omega)]} = 0. \tag{6.61}$$

Defining $d_n \equiv \frac{b_n}{n}$, we find

$$iN\omega\frac{\partial d_n}{\partial n} - \frac{N}{n^2}\log(\varphi) + \frac{N}{n}\frac{\varphi'}{\varphi}\frac{\partial a_n}{\partial n}\omega = 0. \tag{6.62}$$

$$\Rightarrow \frac{\varphi'(a_n\omega)\omega}{\varphi(a_n\omega)} = \frac{\log(\varphi(a_n\omega))}{n\frac{\partial a_n}{\partial n}} - i\omega\frac{\partial d_n}{\partial n}\frac{n}{\frac{\partial a_n}{\partial n}}. \tag{6.63}$$

Multiplying both sides by a_n and defining $\tilde{\omega} \equiv a_n\omega$, we find that

$$\frac{\varphi'(\tilde{\omega})\tilde{\omega}}{\varphi(\tilde{\omega})} - \log(\varphi(\tilde{\omega}))\frac{a_n}{n\frac{\partial a_n}{\partial n}} + i\tilde{\omega}\frac{\partial d_n}{\partial n}\frac{n}{\frac{\partial a_n}{\partial n}} = 0. \tag{6.64}$$

Since this equation should hold (with the same function $\varphi(\tilde{\omega})$) as we vary n, we expect that $\frac{a_n}{n\frac{\partial a_n}{\partial n}}$ and $\frac{\partial d_n}{\partial n}\frac{n}{\frac{\partial a_n}{\partial n}}$ should be nearly independent of n for large values of n. The equation for $\varphi(\tilde{\omega})$ then takes the mathematical structure

$$\frac{\varphi'}{\varphi} - \frac{C_1\log(\varphi(\tilde{\omega}))}{\tilde{\omega}} = iC_2, \tag{6.65}$$

with C_1, C_2 (real) constants (note that by contrast, φ is a complex function of a real variable). We may represent $\varphi = re^{i\theta}$, and obtain two equations for *real* functions of a real variable:

$$\frac{r'}{r} - \frac{C_1 \log(r)}{\tilde{\omega}} = 0, \tag{6.66}$$

$$\theta' - \frac{C_1 \theta}{\tilde{\omega}} = C_2. \tag{6.67}$$

Defining $u = \log(r)$, we find that

$$u' - \frac{C_1}{\tilde{\omega}} u = 0 \tag{6.68}$$

(where u is also a real function of a real variable).

Therefore, the equations for both u and θ take identical form of a simple, linear ODE, albeit with that of θ being *inhomogeneous*. We may follow the general approach of solving the homogenous equation and guessing a particular solution for the inhomogeneous equation (alternatively, we may note that this is a Cauchy–Euler differential equation and can readily be converted into an ODE equation with constant coefficients using the transformation $t \equiv \log(x)$).

We can write the equation for u in the form

$$(\log[u])' = \frac{C_1}{\tilde{\omega}}. \tag{6.69}$$

Upon integrating we find that

$$\log(u) = C_1 \log|\tilde{\omega}| + const \Rightarrow u = a|\tilde{\omega}|^{C_1}, \tag{6.70}$$

where it is important to note that the integration constant and the constant a are real. Since the homogeneous equation corresponding to Eq. (6.67) is identical to Eq. (6.68), its general solution is

$$\theta = b|\tilde{\omega}|^{C_1}, \tag{6.71}$$

with b a real constant. Guessing a particular solution to the inhomogeneous equation in the form $u = D\tilde{\omega}$, leads to

$$D - C_1 D = C_2 \implies D = \frac{C_2}{1 - C_1}, \tag{6.72}$$

with D *real*.

As long as $C_1 \neq 1$, we found a solution! Putting the solutions for r and θ together, the general solution to Eq. (6.65) thus takes the form

$$\varphi(\tilde{\omega}) = e^{A|\tilde{\omega}|^{C_1} + iD\tilde{\omega}}, \tag{6.73}$$

with A a *complex* constant. Importantly, note that since Eq. (6.65) is ill-defined at $\omega = 0$, the constant A may change as ω changes sign – we will shortly see that this indeed is the case (and similarly see that the constant D does not change at $\omega = 0$).

In the case $C_1 = 1$, we can guess a solution of the form $\theta = D\tilde{\omega}\log(|\tilde{\omega}|)$ to Eq. (6.67), and find that

$$D\log(|\tilde{\omega}|) + D - D\log(|\tilde{\omega}|) = C_2, \tag{6.74}$$

hence we have a solution when $D = C_2$ (a *real* constant), which leads to the following form for φ:

$$\varphi(\tilde{\omega}) = e^{A|\tilde{\omega}| + iC_2\tilde{\omega}\log(|\tilde{\omega}|)}. \tag{6.75}$$

Also here, A is a complex constant which may change sign at $\omega = 0$.

Going back to Eq. (6.64), we can also get the approximate scaling for the coefficients (for any value of C_1):

$$\frac{a_n}{n\frac{\partial a_n}{\partial n}} \approx C_1 \implies C_1\frac{\partial \log(a_n)}{\partial n} \approx 1/n. \tag{6.76}$$

This implies that

$$\log(a_n) \approx \frac{1}{C_1}\log(n) + constant \implies a_n \propto n^{1/C_1}. \tag{6.77}$$

Similarly,

$$\frac{\partial d_n}{\partial n}\frac{n}{\frac{\partial a_n}{\partial n}} = -C_2, \tag{6.78}$$

hence,

$$\frac{\partial d_n}{\partial n} \propto n^{1/C_1 - 2}. \tag{6.79}$$

Therefore,

$$d_n = C_3 n^{1/C_1 - 1} + C_4 \implies b_n = C_3 n^{1/C_1} + C_4 n, \tag{6.80}$$

where we relied on our previous definition $d_n = b_n/n$. The first term will become a constant when we divide by the term $a_n \propto n^{1/C_1}$ of Eq. (6.54), leading to a simple shift of the resulting distribution. Upon dividing by the term a_n, the second term will vanish for large n when $C_1 < 1$, as we anticipated based on the results of the previous section, on positive Lévy stable distributions. We shall soon see that the case $C_1 > 1$ corresponds to the case of a variable with finite mean, in which case the $C_4 n$ term will be associated with centering of the scaled variable by subtracting their mean, as in the standard CLT.

A word of caution The constraint imposed by the RG approach is insufficient in pinning down the scaling factor a_n precisely. Really, all we know is that $\lim_{n\to\infty}\frac{a_n}{n\frac{\partial a_n}{\partial n}}$ should tend to a constant. In the above, we solved the ODE resulting from equating this term to a constant, but it is easy to see that modulating this power-law by, e.g., logarithmic corrections (or powers thereof) would also satisfy the RG requirement. Indeed, Problem 6.4 deals with a case where $\mu = 2$ but the scaling of a_n contains logarithmic corrections to the expected $\sqrt{n}$ scaling. (As we shall shortly see, in the case

$\mu < 2$ if the original distribution has a power-law tail, the scaling of the coefficients a_n will not contain such logarithmic corrections). Similarly, care should be taken in interpreting the power-law scaling of the coefficients b_n, which we discuss next, as well as their counterparts in the "extreme value distributions" later on. Nevertheless, in many applications knowing the leading order dependence of the coefficients on n suffices, which is adequately captured by the RG-inspired approach.

6.8.1 General Formula for the Characteristic Function of Lévy Stable Distributions

According to Eq. (6.73), the general formula for the characteristic function of $p(\xi_n)$ for $C_1 \neq 1$ is

$$\varphi(\omega) = e^{A|\omega|^{C_1} + iD\omega}. \tag{6.81}$$

The requirement that the inverse Fourier transform of φ is a probability distribution imposes that $\varphi(-\omega) = \varphi^*(\omega)$. Therefore, the characteristic function takes the form

$$\varphi = \begin{cases} e^{A\omega^{C_1} + iD\omega}, & \omega > 0, \\ e^{A^*|\omega|^{C_1} + iD\omega}, & \omega < 0. \end{cases} \tag{6.82}$$

(As noted previously, the value of A in Eq. (6.81) was indeed "allowed" to change at $\omega = 0$.) The D term is associated with a trivial shift of the distribution (related to the linear scaling of b_n) and can be eliminated. We will therefore not consider it in the following. The case of $C_1 = 1$ will be considered in the next section.

Eq. (6.82) may be rewritten as

$$\varphi = e^{-a|\omega|^{\mu}[1 - i\beta sign(\omega) tan(\frac{\pi\mu}{2})]}, \tag{6.83}$$

with a and β real constants and where clearly $\mu = C_1$. The asymmetry of the distribution is determined by β. For this representation of φ, we will now show that $-1 \leq \beta \leq 1$, and that $\beta = 1$ corresponds to the case we have studied earlier of a positive distribution. Similarly, $\beta = -1$ corresponds to a strictly negative distribution, while $\beta = 0$ corresponds to a symmetric one. This formula will not be valid for $\mu = 1$, a case that we will discuss in the next section.

Consider, $p(x)$ which decays, for $x > x^*$, as

$$p(x) = \frac{A_+}{x^{1+\mu}}, \tag{6.84}$$

with $0 < \mu < 1$. We will explicitly assume that the function decays sufficiently fast for $x \to -\infty$, and later generalize to the case of both a right and left power-law tail. We shall now derive yet another Tauberian relation, finding the form of the Fourier transform of $p(x)$ near the origin in terms of the tail of the distribution. For small, positive ω we find that

$$\Phi(\omega) \equiv \int_{x^*}^{\infty} \frac{A_+}{x^{1+\mu}} e^{i\omega x} dx = A_+ \int_{x^*\omega}^{\infty} \frac{e^{im}}{m^{1+\mu}} dm \frac{\omega^{1+\mu}}{\omega}, \tag{6.85}$$

where we substituted $m = \omega x$. Evaluating the integral on the RHS by parts we obtain

$$\Phi(\omega) = A_+ \left[-\omega^\mu \frac{m^{-\mu}}{\mu} e^{im} \Big|_{x^*\omega}^{\infty} \right] + A_+ \omega^\mu \int_{x^*\omega}^{\infty} \frac{im^{-\mu}}{\mu} e^{im} dm. \tag{6.86}$$

For $\mu < 1$, we may approximate the integral by replacing the lower limit of integration by 0, to find that

$$\Phi(\omega) \approx A_+ \frac{x^{*-\mu}}{\mu} e^{ix^*\omega} + A_+ \frac{\omega^\mu}{\mu} i \int_0^{\infty} \frac{e^{im}}{m^\mu} dm \tag{6.87}$$

and $\int_0^{\infty} \frac{e^{im}}{m^\mu} dm = i\Gamma(1 - \mu)e^{-i\frac{\pi}{2}\mu}$ (this can easily be evaluated using contour integration). Thus, for small, positive ω we have

$$\Phi(\omega) \approx C_0 - C_+\omega^\mu, \tag{6.88}$$

with C_0, C_+ constants. $\Phi(\omega)$ is not the characteristic function, since the lower limit of the integration in Eq. (6.85) is x^*. However, due to our assumption that the left tail decays fast, we can bound the rest of the integral to be at most linear in ω for small ω (up to a constant), in a similar way to the one used in our discussion of Laplace transforms (similar results will hold if the left tail is power-law albeit with a larger power than the right tail). Therefore, the characteristic function near the origin is approximated by

$$\varphi(\omega) \approx 1 - C_+\omega^\mu, \tag{6.89}$$

with $C_+ = A_+ \frac{\Gamma(1-\mu)e^{-i\frac{\pi}{2}\mu}}{\mu}$. We have

$$\frac{\text{Im}(C_+)}{\text{Re}(C_+)} = -tan\left(\frac{\pi}{2}\mu\right) \Rightarrow C_+ = a\left[1 - i \cdot tan\left(\frac{\pi}{2}\mu\right)\right], \tag{6.90}$$

with a a real coefficient. This corresponds to $\beta = 1$ in our previous representation. If we similarly look at a distribution with a left tail, a similar analysis leads to the same form of Eq. (6.89) albeit with $C_- = a[1 + i \cdot tan(\frac{\pi}{2}\mu)]$ (and a real), corresponding to $\beta = -1$ in Eq. (6.83). In the general case where the function has both a left and right power-law tail $\frac{A_\pm}{x^{1+\mu}}$, we obtain the expression (for positive ω)

$$\varphi(\omega) \approx 1 - \tilde{C}\omega^\mu(A_+e^{-i\mu\frac{\pi}{2}} + A_-e^{i\mu\frac{\pi}{2}}), \tag{6.91}$$

with $\tilde{C} \equiv \frac{\Gamma(1-\mu)}{\mu}$.

We can write this as $\varphi(\omega) \approx 1 - C\omega^\mu$, where now we have

$$\frac{\text{Im}(C)}{\text{Re}(C)} = \frac{-sin(\frac{\pi}{2}\mu)}{cos(\frac{\pi}{2}\mu)}\left(\frac{A_+ - A_-}{A_+ + A_-}\right) = -tan\left(\frac{\pi}{2}\mu\right)\beta, \tag{6.92}$$

with β defined as

$$\beta = \frac{A_+ - A_-}{A_+ + A_-}. \tag{6.93}$$

This clarifies the notation of Eq. (6.83), and why β is restricted to the range $[-1, 1]$.

A tail of tales – and black swans It is interesting to note that unlike the case of finite variance, here the limiting distribution depends only on A_+ and A_-: the tails of the original distribution. The behavior is only dominated by these tails – even if the power-law behavior only sets in at large values of x! This also brings us to the concept of a "black swan": scenarios in which rare events – the probability of which is determined by the tails of the distribution – may have dramatic consequences. Here, such events dominate the sums. Later, we will see physical examples of such events in the context of anomalous diffusion and Lévy flights. For a popular discussion of black swans and their significance, see Taleb (2007). Finally, Problem 6.9 studies the tails of the Lévy distributions themselves – perhaps not surprisingly, these also manifest a power-law tail. Problem 6.14 shows that for sufficiently heavy-tailed distribution, anomalously large sums will always be dominated by a *single* large event!

What about the case where $1 < \mu < 2$? Following the same logic – with an additional integration by parts - we find that the form of Eq. (6.83)(with $|\beta| \le 1$) is still intact also for $1 < \mu < 2$ (see Problem 6.7). Note that the linear term will drop out due to the shift of Eq. (6.54). It is also worth mentioning that the asymmetry term vanishes as $\mu \to 2$: In the case of finite variance we always obtain a symmetric Gaussian, i.e., we become insensitive to the asymmetry of the original distribution. Finally, one may wonder what happens when $\mu > 2$? Problem 6.10 shows that in this case although Eq. (6.81) is a formal solution of the RG-inspired procedure, it does not correspond to a probability distribution.

Special Cases
$\mu = 1/2, \beta = 1$: *Lévy Distribution*
Consider the Lévy distribution of Eq. (6.46):

$$p(x) = \sqrt{\frac{C}{2\pi}} \frac{e^{-\frac{C}{2x}}}{(x)^{3/2}} \quad (x \ge 0). \tag{6.94}$$

The Fourier transform of $p(x)$ for $\omega > 0$ is

$$\varphi(\omega) = e^{-\sqrt{-2iC\omega}}, \tag{6.95}$$

which indeed corresponds to $\tan(\frac{\pi}{2}\frac{1}{2}) = 1 \to \beta = 1$. The Laplace transform is

$$L(s) = e^{-\sqrt{2Cs}}, \tag{6.96}$$

as we have previously seen when discussing *positive* Lévy stable distributions.

$\mu = 1$: *Cauchy Distribution and More*
We have previously analyzed the case $\mu = 1, \beta = 0$, corresponding to the Cauchy distribution. In the general case $\mu = 1$ and $\beta \ne 0$, we have seen that the general form of the characteristic function is, according to Eq. (6.75),

$$\varphi(\omega) = e^{A|\omega| + iD\omega \log(|\omega|)}. \tag{6.97}$$

Repeating the logic we used before to establish the coefficients D for $\omega > 0$ and $\omega < 0$, based on the power-law tails of the distributions (which in this case fall off like $1/x^2$) leads to

$$\varphi(\omega) = e^{-a|\omega|[1-i\beta sign(\omega)\phi]}; \quad \phi = -\frac{2}{\pi}\log|\omega|, \tag{6.98}$$

with a and β real constants. This is the only exception to the form of Eq. (6.83). It remains to be shown why $\beta = 1$ corresponds to the case of a strictly positive distribution (which would thus justify the $\frac{2}{\pi}$ factor in the definition of ϕ). To see this, note that the logic following up to Eq. (6.86) is still intact for the case $\mu = 1$. However, we can no longer replace the lower limit of integration by 0 in Eq. (6.87). The real part of the integral can be evaluated by parts, leading to a $-\log(\omega)$ divergence. The imaginary part of the integral does not suffer from such a divergence and can be approximated by replacing the lower limit of integration with 0. Using $\int_0^\infty \frac{\sin(x)}{x}dx = \pi/2$, we find that for small, positive ω,

$$\Phi(\omega) \approx c[\pi/2 + i\log(\omega)], \tag{6.99}$$

with c a real number, corresponding indeed to the form of Eq. (6.98) with $\beta = 1$.

6.9   Exploring the Stable Distributions Numerically

Since we defined the stable distributions in terms of their Fourier transforms, it is very easy to find them numerically by calculating the inverse Fourier transform. Here is a short code that finds the stable distribution for $\mu = 0.5$ and $\beta = 0.5$ (i.e., corresponding to a distribution with right and left power-law tails with asymmetric magnitudes). It is easy to check that if β exceeds one, the inverse Fourier transform leads to a function with negative values – and hence does not correspond to a probability distribution – as we anticipated.

```
mu=0.5; B=0.5; dt=0.001;

w_vec=-10000:dt:10000;
for indx=1:length(w_vec)
    w=w_vec(indx);
    f(indx)=exp(-abs(w)^mu/2*(1-i*B*tan(pi*mu/2)*sign(w)));
end;

x_vec=-5:0.1:5;
for indx=1:length(x_vec)
    x=x_vec(indx);
    y(indx)=0;
    for tmp=1:length(f)
    y(indx)=y(indx)+exp(-i*w_vec(tmp)*x)*f(tmp)*dt;
    end;
end;
y=y/(2*pi);
plot(x_vec,y);
```

Fig. 6.6 shows the result.

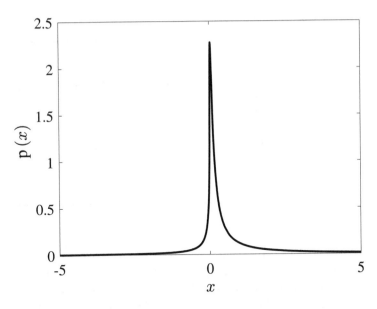

Figure 6.6 Lévy stable distribution for $\mu = 0.5$ and asymmetry $\beta = 0.5$, obtained by numerically inverting Eq. (6.83).

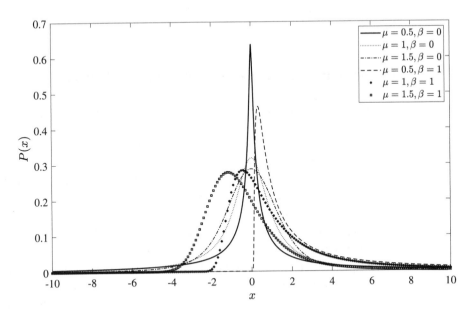

Figure 6.7 Lévy stable distributions for μ ranging from 0.5 to 1.5 and asymmetry β ranging from 0 to 1.

There are much better numerical methods to evaluate the Lévy stable distributions (http://math.bu.edu/people/mveillet/html/alphastablepub.html). Fig. 6.7 shows examples for Lévy distributions obtained using such a method. Note that for $\mu < 1$ and $\beta = 1$ the support of the stable distribution is strictly positive, while this is not the

Table 6.1 Lévy stable distributions and their scaling

Tail $\propto 1/x^{1+\mu}$	Variance	Mean	Scaling	Fourier representation (after scaling)
$\mu > 2$	Finite, σ^2	Finite, μ	$\frac{S_n - \mu n}{\sigma\sqrt{n}}$	Gaussian
$1 < \mu < 2$	Infinite	Finite, μ	$\frac{S_n - \mu n}{n^{1/\mu}}$	Eq. (6.83)
$0 < \mu < 1$	Infinite	Infinite	$\frac{S_n}{n^{1/\mu}}$	Eq. (6.83)
$\mu = 1$	Infinite	Infinite	$\frac{S_n}{n}$	Eq. (6.98)

case for $\mu > 1$. This is expected, since for $\mu \leq 1$ we have seen that we can set the shift b_n arbitrarily to zero and still converge to the Lévy stable distribution. Since the initial distribution is strictly positive, the scaled sum will also be strictly positive, and hence the result for the limiting distribution follows. Note that this argument fails for $\mu > 1$.

Table 6.1 summarizes some of the results derived so far in this chapter (note that the case $\mu = 2$ is not included here – it is studied in Problem 6.4).

6.10   RG-Inspired Approach for Extreme Value Distributions

Consider the maximum of n variables drawn from some distribution $p(x)$, characterized by a cumulative distribution $C(x)$ (i.e., $C(x)$ is the probability for the variable to be smaller than x). It vanishes for $x \to -\infty$ and approaches 1 as $x \to \infty$. We will now find the behavior of the maximum for large n, that will turn out to also follow universal statistics – much like in the case of the GCLT – that depend on the tails of $p(x)$. This was discovered by Ronald Fisher and L. H. C. Tippett (1928), motivated by an attempt to characterize the distribution of strengths of cotton fibers and has since found a plethora of diverse applications as reviewed in Fortin and Clusel (2015). Our approach will be reminiscent of (yet distinct from) that of Fisher and Tippett, and will in fact closely follow the RG-inspired approach we used for deriving the Lévy stable distributions, albeit with the *cumulative* distribution function replacing the role of the *characteristic* function – for reasons that will shortly become clear. Note that the derivation that will be presented below only relies on the fact that taking the maximum through different procedures should lead to the same result, as we did in deriving the GCLT. This self-similarity approach is in the spirit of RG approaches, but not an RG approach *par excellence* as utilized in, e.g., Bertin and Györgyi (2010); Calvo *et al.* (2012); Manzato *et al.* (2012); Györgyi *et al.* (2008, 2010); Bazant (2000). For a discussion on the similarities and distinctions between the different approaches and a list of relevant references, see Amir (2020).

Extreme values We will be interested in the maximum (or equivalently, minimum) of a *large* number of variables. By nature, this (rare) random event is an outlier – the largest or smallest over many trials (assumed here to be independent). Indeed, the results are often applied to problems where the extreme events matter – what should be the height of a dam? What is the chance of observing an earthquake or tsunami of a given magnitude? As in the original application (assessing the statistics of the strength of cotton fibers), these ideas are often invoked in the context of materials failure, see, for example, Le, Bazant, and Bazant (2011, 2009). For these reasons insurance companies should be highly interested in this chapter.

To begin, we define

$$X_n \equiv \max(x_1, x_2, \ldots, x_n), \tag{6.100}$$

where $x_1, \ldots, x_n$ are again i.i.d. variables.

Since we have

$$\text{Prob}[X_n < x] = \text{Prob}[x_1 < x]\text{Prob}[x_2 < x]\ldots \text{Prob}[x_n < x] = C^n(x), \tag{6.101}$$

it is natural to work with the cumulative distribution when dealing with extreme value statistics, akin to the role which the characteristic function played in the previous section. Clearly, it is easy to convert the question of the *minimum* of n variables to one related to the maximum, if we define $\tilde{p}(x) = p(-x)$.

Before proceeding to the general analysis, which will yield three distinct universality classes (corresponding to the Gumbel, Weibull, and Fréchet distributions), we will exemplify the behavior of each class on a particular example.

6.10.1 Example I: The Gumbel Distribution

Consider the distribution

$$p(x) = e^{-x}. \tag{6.102}$$

Its cumulative is

$$C(x) = 1 - e^{-x}. \tag{6.103}$$

The cumulative distribution for the maximum of n variables is therefore

$$G(x) = (1 - e^{-x})^n \approx e^{-ne^{-x}} = e^{-e^{-(x-x_0)}}, \tag{6.104}$$

with $x_0 \equiv \log(n)$. Upon defining a shifted variable $\tilde{x} = x - \log(n)$, the cumulative of the distribution will be approximately $e^{-e^{-\tilde{x}}}$. This is an example of the *Gumbel distribution* (named after Emil Julius Gumbel). The general form of its cumulative is

$$G(x) = e^{-e^{-(ax+b)}}. \tag{6.105}$$

Taking the derivative of Eq. (6.104) to find the probability distribution for the maximum, we find that

$$p_n(x) = e^{-e^{-(x-x_0)}} e^{-(x-x_0)} \qquad (6.106)$$

(where the n dependence enters only via x_0). Denoting $l \equiv e^{-(x-x_0)}$, we have

$$p(x) = e^{-l} l, \qquad (6.107)$$

and taking the derivative with respect to l we find that the distribution is peaked at $x = \log(n)$. It is easy to see that its width is of order unity. We can now revisit the approximation we made in Eq. (6.104), and check its validity.

We have seen before (in Section 2.5) that making the approximation

$$(1 - x/n)^n \approx e^{-x} \qquad (6.108)$$

is valid under the condition $x \ll \sqrt{n}$. In our case, this implies that

$$e^{-x} n \ll \sqrt{n}. \qquad (6.109)$$

At the peak of the distribution, we have $e^{-x} n = 1$, and the approximation is valid for $n \gg 1$. From Eq. (6.109) we see that the approximation we used would break down when we take x to be sufficiently smaller than $\log(n)$. Defining $x = \log(n) - \delta x$, we see that the value of δx for which the approximation fails obeys $e^{-\delta x} = O(\sqrt{n})$, hence $\delta x = O(\log(\sqrt{n}))$. Since, as we saw earlier, the width of the distribution is of order unity, this implies that for large n the Gumbel distribution would approximate the exact solution well, failing only sufficiently far in the (inner) tail where the probability distribution is vanishingly small. However, a note of caution is in place: The logarithmic dependence we found signals a very slow convergence to the limiting form. This is also true in the case where the distribution $p(x)$ is Gaussian, as was already noted in original work of Fisher and Tippett (1928).

We will soon show that for the maximum of a large number of variables drawn from *any* distribution with a sufficiently fast decaying tail, the Gumbel distribution arises.

6.10.2   Example II: The Weibull Distribution

Consider the *minimum* of the same distribution we had in the previous example. The same logic would give us that

$$\text{Prob}[\min(x_1,...x_n) > x] = \text{Prob}[x_1 > x]\text{Prob}[x_2 > x]..\text{Prob}[x_n > x] = e^{-nx}. \qquad (6.110)$$

We may define a new, scaled variable as

$$\tilde{x} = \min(x_1, \ldots x_n)n \qquad (6.111)$$

to find that $P(\tilde{x}) = e^{-\tilde{x}}$ for the above example. Therefore, by "zooming into" the cutoff of the distribution (zero for this example) we converge to an n-independent limiting distribution.

This is an example of the Weibull distribution (named after Waloddi Weibull), which occurs when the variable is bounded (e.g., in this case the variable is never negative). As we shall see, the limiting cumulative distribution for the maximum of n variables with distribution bounded by x^* will be

$$G(x) = \begin{cases} e^{-a(x^*-x)^{1/|\alpha|}}, & x \leq x^*, \\ 0, & x > x^*. \end{cases} \tag{6.112}$$

The behavior of the original distribution $p(x)$ near the cutoff x^* is important, and determines the exponent α. We will also see that to leading order the scaling in this case takes a power-law form in n (linear in the above example). The form of the power-law scaling will similarly depend on the behavior of the original distribution near its cutoff.

6.10.3 Example III: The Fréchet Distribution

The final example belongs to the third possible universality class, corresponding to variables with a power-law tail.

If at large x we have

$$p(x) = \frac{A_+}{(x - B)^{1+\mu}}, \tag{6.113}$$

then the cumulative distribution is

$$C(x) = 1 - \frac{A_+}{\mu(x - B)^{\mu}}. \tag{6.114}$$

Therefore, taking it to a large power n we find that

$$C^n(x) \approx e^{-\frac{A_+ n}{\mu(x-B)^{\mu}}}. \tag{6.115}$$

Upon a linear scaling,

$$\tilde{x} = \frac{x - B}{n^{1/\mu}}, \tag{6.116}$$

we find that

$$G(\tilde{x}) = e^{-a\tilde{x}^{-\mu}} \tag{6.117}$$

(where the coefficient a does not depend on n). This limiting distribution is known as the Fréchet distribution, named after Maurice René Fréchet. Importantly, we see that in this case the width of the distribution (before the linear scaling) increases with n as a power-law $n^{1/\mu}$. For $\mu \leq 1$, this is precisely the same scaling we derived for the *sum* of n variables drawn from this heavy-tailed distribution! This elucidates why in the scenario $\mu \leq 1$ we obtained *Lévy flights*, where the sum was dominated by rare events no matter how large n was. This is also related to the so-called Single Big Jump Principle, studied in Problem 6.14, showing that for large N the *tail* of the distribution of the sum for $\mu < 2$ will be dominated, remarkably, by a *single* jump

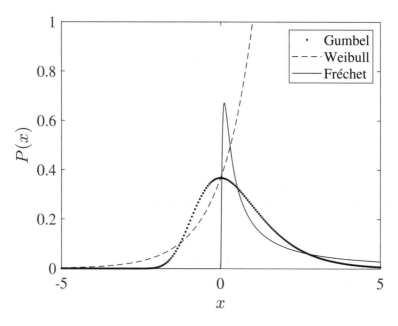

Figure 6.8 Examples of extreme value distributions.

(note, however, that this different result is obtained only when conditioning on the sum being anomalously large – such that we are in the tail of the distribution). This idea has been recently shown to pertain to a broader class of scenarios in physics, extending the results for i.i.d. variables (see Vezzani, Barkai, and Burioni (2019); Wang *et al.* (2019) and references therein), as well as applications in finance (Filiasi *et al.* 2014).

The extreme value distributions (i.e., the limiting distributions) corresponding to the three examples above are shown in Fig. 6.8 (note that for convenience of comparison, the Weibull example is shown for the *maximum* of variables drawn from an exponentially decaying distribution).

We shall now show that these three cases can be derived in a general framework, using a similar approach to the one we used earlier.

6.10.4   General Form for Extreme Value Distributions

We will next follow similar logic to the RG-inspired calculation used to find the form of the characteristic functions in the GCLT to find all possible extreme value distributions. By itself, our analysis will not reveal the "basin of attraction" of each universality class, nor will we find the *precise* scaling of the coefficients a_n and b_n. These require work beyond the basic RG-inspired calculation presented here.

As before, let us assume that some scaling coefficients a_n, b_n exist such that when we define

$$\xi_n \equiv \frac{X_n - b_n}{a_n}, \tag{6.118}$$

the following limit exists:

$$\lim_{n \to \infty} \text{Prob}[\xi_n = \xi] = g(\xi) \tag{6.119}$$

(note that this limit is not unique: we can always shift and rescale by a constant). The function g is called the Extreme Value Distribution (EVD).

Scaling Note that in various references the scaling coefficients are introduced such that $G_n[a_n x + b_n]$ approaches a limiting distribution, where G_n is the cumulative distribution for a sum of n i.i.d. variables. This is equivalent to the form of scaling introduced above. Beware that in some sources, however, a_n and b_n are flipped with respect to the notation used here, e.g., Vivo (2015).

This implies that $p(X_n) \approx a_n^{-1} g\left(\frac{X_n - b_n}{a_n}\right)$ and the cumulative is given by $G\left(\frac{X_n - b_n}{a_n}\right)$. By the same logic we used before, we know that $G^m\left(\frac{X_n - b_n}{a_n}\right)$ depends only on the quantity $N = n \cdot m$. Therefore, we have

$$\frac{\partial}{\partial n}\left(G^{N/n}\left(\frac{X_n - b_n}{a_n}\right)\right) = 0. \tag{6.120}$$

(Note that here X_n is the random variable we are interested in: Hence while derivatives of the *coefficients* a_n, b_n appear, a derivative of X_n is not defined and does not appear). This leads to

$$\frac{\partial}{\partial n}\left(\frac{N}{n} \log\left(G\left(\frac{X_n - b_n}{a_n}\right)\right)\right) = 0, \tag{6.121}$$

from which we find that

$$-\frac{N}{n^2} \log G + \frac{N}{n} \frac{G'}{G}\left[-\frac{\partial}{\partial n}\left(\frac{b_n}{a_n}\right) - \frac{X_n}{a_n^2} \frac{\partial a_n}{\partial n}\right] = 0. \tag{6.122}$$

Upon defining a new random variable $\tilde{x} \equiv \frac{X_n - b_n}{a_n}$ (as in Eq. (6.118)), the equation can be rewritten as

$$[\log(-\log G(\tilde{x}))]' = -\left[\frac{n}{a_n} \frac{\partial b_n}{\partial n} + \frac{n}{a_n} \frac{\partial a_n}{\partial n} \tilde{x}\right]^{-1}. \tag{6.123}$$

In order for the RHS to have a sensible limit for large n, we would like to have

$$\lim_{n \to \infty} \frac{n}{a_n} \frac{\partial a_n}{\partial n} = \alpha \tag{6.124}$$

with α constant. Similarly, we have

$$\lim_{n \to \infty} \frac{n}{a_n} \frac{\partial b_n}{\partial n} = \beta \tag{6.125}$$

with β constant.

We shall shortly show that the value of α will dictate to which of the three universality classes we will converge.

Fréchet Distribution

If $\alpha > 0$, we find that to leading order (interpreted in the same subtle way as we previously did for the Lévy stable distributions, as discussed in the box "A word of caution"):

$$a_n \propto n^\alpha. \tag{6.126}$$

Next, requiring that $\frac{n}{a_n}\frac{\partial b_n}{\partial n}$ should be constant implies that $b_n \propto a_n$. This corresponds to a *shift* in the scaled variable, and therefore we can set $b_n = 0$ without loss of generality.

Solving for $G(\tilde{x})$ gives the Fréchet distribution:

$$G(x) = e^{-ax^{-1/\alpha}}. \tag{6.127}$$

Comparing this form with Eq. (6.117), we recognize that $1/\alpha = \mu$.

Weibull Distribution

Similarly, when $\alpha < 0$ we find that to leading order,

$$a_n \propto n^{-|\alpha|}. \tag{6.128}$$

Solving for G gives us the Weibull Distribution,

$$G(x) = e^{a(\beta/|\alpha|-x)^{1/|\alpha|}}, \tag{6.129}$$

with a some constant. We see that for this class, the cumulative distribution equals precisely 1 at $x = \beta/|\alpha|$, implying that the distribution vanishes for x larger than this value (while it is manifestly nonzero for smaller x). Note that this threshold is arbitrary: We can always add a term proportional to a_n to the scaling coefficients b_n, and shift the limiting distribution as desired. Indeed, according to Eqs. (6.128) and (6.125), we have $b_n \approx \hat{x} + \beta a n^{-1/|\alpha|} = \hat{x} + \hat{s}a_n$, with $\hat{x}$ and $\hat{s}$ constants. We conclude that for large n the coefficients b_n are approximately *constant*, and that the coefficients a_n monotonically *decrease* with n. Since for large n the cumulative of the original distribution is well approximated by $G\left(\frac{X_n - b_n}{a_n}\right)$, and since the support of the limiting distribution is bounded from above by $\beta/|\alpha|$, we conclude that the original distribution is also bounded from above by $\lim_{n\to\infty} a_n\beta/|\alpha| + b_n = \hat{x}$. The above relation implies that we are essentially "zooming into" a particular region of the probability distribution as n increases. The distribution of the maximum becomes *narrower* as n increases, and is focused near its upper bound $\hat{x}$.

As mentioned before, the coefficient α is determined by the behavior of the original probability distribution near the cutoff x^* (which we showed above must equal $\hat{x}$). In the particular example discussed earlier $p(x)$ approached a nonzero constant near x^*, hence we found $|\alpha| = 1$. It is straightforward to generalize this to the case where $p(x)$ vanishes near the cutoff as a power-law $(x - x^*)^c$, finding that $1/|\alpha| = c + 1$.

Gumbel Distribution

Finally, consider the case $\alpha = 0$. Given that $\frac{n}{a_n} \frac{\partial b_n}{\partial n}$ is approximately constant, we obtain the Gumbel distribution:

$$G(x) = e^{-e^{-(ax+b)}}. \tag{6.130}$$

In the two previous cases where $\alpha \neq 0$, we found that the leading order behavior of the coefficients a_n, b_n was pinned-down by the RG-inspired approach. This is not the case when the RHS of Eq. (6.124) vanishes. In this case, unfortunately, the leading order of the scaling coefficients a_n is nonuniversal, and therefore cannot be determined from the RG-inspired approach alone. According to Eq. (6.125), the same holds for the scaling coefficients b_n. For $\alpha = 0$ both scaling coefficients a_n and b_n must be determined from the tail of $p(x)$. For example, Problem 6.13 shows that for the Gaussian distribution a particular (but non-unique) choice of scaling coefficients that leads to convergence to a Gumbel distribution is (Fisher and Tippett 1928; Vivo 2015):

$$a_n = 1/b_n; b_n = \sqrt{2\log(n) - \log(4\pi \log(n))}. \tag{6.131}$$

Another word of caution. Throughout the work, the logic of the RG-inspired approach *assumes* we have some scaling coefficients a_n, b_n such that the limit of Eq. (6.119) exists, and draws the (rather strong) constraints on the limiting distributions and scaling coefficients from this condition. But it is not a priori clear that such a scaling is at all possible! Problem 6.15 gives a simple example of a probability distribution where there is no convergence to an extreme value distribution (see also Amir 2020). In fact, general criteria exist which determine whether a particular distribution has a limiting EVD, and if it does, to which of the three universality classes it belongs – see, for example, Vivo (2015).

6.11 Summary

This chapter was a tale of tails – primarily dealing with distributions that are "heavy-tailed" leading to the breakdown of the central limit theorem. We began by exploring sums of i.i.d. random variables, and realized that going to Fourier space is a natural thing to do since convolutions (describing the probability distribution of the sum) turn into products. This led us to define the *characteristic function*. Moreover, we realized that for a large number of variables, we are primarily interested in the form of the characteristic function at small frequencies – which led us to the notion of Tauberian theorems. These relate the small frequency behavior of the characteristic function to the tail of the probability distribution (due to the reciprocity of real and Fourier space). For variables with a positive support, it was effective to repeat this sort of analysis using Laplace rather than Fourier transforms, with a relevant application to slow relaxations in nature. We used a renormalization-group-inspired approach to find all possible limiting distributions of the sums, leading us to a generalization of the central limit theorem to heavy-tailed distributions, involving the Lévy stable distributions. Finally, we used a similar approach to study the similarly universal behavior of the *maximum* of a large number of i.i.d. variables (extreme value distributions), in which case the cumulative distribution played the part previously taken by the characteristic function. We showed that the family of extreme value distributions consist of three classes: Gumbel, Weibull, and Fréchet.

For further reading Taleb (2007) is a popular (non-technical text) on the relevance of rare events ("black swans") to our world. Feller (2008) provides a rigorous treatment of many of the results derived in this chapter. Sornette (2006), Hughes (1996), and Paul and Baschnagel (2013) provide a more extensive discussion of many of the topics discussed in this chapter.

6.12   Exercises

6.1   Pearson's Problem

The great statistician (and polymath) Karl Pearson posted a problem in the magazine *Nature* in 1905 (see below). Let us help him solve it!

> **The Problem of the Random Walk.**
>
> CAN any of your readers refer me to a work wherein I should find a solution of the following problem, or failing the knowledge of any existing solution provide me with an original one? I should be extremely grateful for aid in the matter.
>
> A man starts from a point O and walks *l* yards in a straight line; he then turns through any angle whatever and walks another *l* yards in a second straight line. He repeats this process *n* times. I require the probability that after these *n* stretches he is at a distance between *r* and *r* + δ*r* from his starting point, O.
>
> The problem is one of considerable interest, but I have only succeeded in obtaining an integrated solution for *two* stretches. I think, however, that a solution ought to be found, if only in the form of a series in powers of 1/*n*, when *n* is large.   KARL PEARSON.
> The Gables, East Ilsley, Berks.

Consider an isotropic random walk in 2D, with i.i.d. displacements of length *l* described by the PDF

$$p(x, y) = \frac{\delta(r - l)}{2\pi l}, \quad r = \sqrt{x^2 + y^2}. \tag{6.132}$$

(a) Assuming that the walker starts at the origin, find the characteristic function of $p(\vec{x})$ and use this to determine the PDF of the position of the walker after n steps as an integral expression. *Hint: it may be helpful to use polar coordinates.*

(b) By taking appropriate leading-order terms, derive analytically the asymptotic form of $p(n)$ as $n \to \infty$ for $r \sim \sqrt{n}$.

(c) How does the width of the central region scale with n?

6.2 Random Walks in 3D and the Central Region

Consider an isotropic random walk in 3D, with i.i.d. displacements of length l described by the PDF

$$p(\vec{x}) = \frac{\delta(r - l)}{4\pi l^2}, \quad r = |\vec{x}|.$$

(a) Assuming that the walker starts at the origin, find the characteristic function of $p(\vec{x})$ and use this to determine the PDF of the position of the walker after n steps as an integral expression. *Hint: use spherical coordinates.*

(b) By taking appropriate leading-order terms, derive analytically the asymptotic form of p_n as $n \to \infty$ for $r \sim \sqrt{n}$.

(c) How does the width of the central region scale with n?

This problem was adapted by Felix Wong based on a similar problem by Martin Bazant.

6.3 Large Deviation Function for Sum of Random Variables

Consider a sum of N i.i.d. variables that (i) Take the values $0, 1$ with equal probability. (ii) Take the values $0, 1$ with probabilites $1/3$ and $2/3$, respectively.

(a) Show that in both cases you may write the probability distribution as $p_N(x) \approx e^{-NI(x/N)}$, identifying the function I for these two cases. Note: do not limit yourself to the regime where the sum is Gaussian.

(b) Use this approach to find the width of the central region for these two examples.

Problem credit: Prof. Eli Barkai.

6.4 Long Range Interactions*

The electric potential created by a charge a distance r away from it is Q/r, and is additive (i.e., for two charges it would be $Q_1/r_1 + Q_2/r_2$).

(a) Consider a large $N \times N$ ordered array of charges in two dimensions

$$+ - + - + - + - + -$$
$$- + - + - + - + - +$$
$$+ - + - + - + - + -$$

What is the potential felt by one of the positive charges at the center of the square? Does a limit exist as $N \to \infty$?

(b) Consider now N charges, each with a randomly chosen charge $+/-$, distributed randomly and uniformly in a disk with radius $\sqrt{N}$. What is (approximately) the probability distribution for the potential at the center of the disk? What happens as $N \to \infty$?

6.5 Power-law Tailed Distributions Again

Consider the sum of N variables drawn from the distribution $p(x) = c/(1 + x^4)$.

(a) What is the range over which the distribution of the sum is approximately Gaussian?

(b)   What is the probability distribution of the sum for large values of x away from that region?

6.6   Simulating the Statistics of Power-law Tailed Distributions

Consider a random variable X with the power-law distribution $p(x) \propto \frac{1}{x^{1+\mu}}$ for $x \in [x_{\min}, \infty)$, where $p(x)$ vanishes for $x < x_{\min}$. For each part below, consider the following three cases of μ: (1) finite mean and finite variance; (2) finite mean and infinite variance; (3) infinite mean and infinite variance.

(a)   Numerically draw $N = 10^4$ numbers from this distribution. *Hint:* You are given a random number y chosen uniformly in $[0, 1]$. What is $f(y)$ such that the distribution of $x = f(y)$ will follow the desired distribution $p(x)$? Plot the running sum $Z(m) = \sum_{i=1}^{m} X_i$, for different values of $m \leq N$, for cases (1) to (3).

(b)   By repeating the procedure a sufficient number of times, calculate the probability distribution of $Z(m)$ for $m = 10^3, 10^4$, and 10^5, for cases (1) to (3). For each case, show how the three distributions can be collapsed using the appropriate scaling.

(c)   Repeat the simulations in parts (a) and (b) for cases (1) to (3), but consider the maximum $M(m) = \max(X_1, \ldots, X_m)$ instead. Run the simulation for $m = 10^3$, 10^4, and 10^5, and show how the distributions can be collapsed using the appropriate scaling.

(d)   Consider $\rho = \frac{\max(X_1, \ldots, X_m)}{\sum_{i=1}^{m} X_i}$ which quantifies the dominance of the largest jump in the process. By repeating the procedure $N = 100$ times, calculate the distribution of ρ for case (1) to (3). Extend N to $1,000$, $10,000$, and $100,000$. Is the distribution dependent on N for each case?

6.7   Finite Mean, Diverging Variance

(a)   Let p be a distribution with power-law tails $p(x) \sim A_{\pm} |x|^{-(1+\mu)}$ as $x \to \pm\infty$ for $1 < \mu < 2$. Find the form of the characteristic function for small ω.

(b)   Now consider the scaled sum of n such random variables, $\xi_n = \frac{S_n - n\langle x \rangle}{n^{1/\mu}}$. Show that as $n \to \infty$, the characteristic function of ξ_n converges to Eq. (6.83) and identify β in terms of $A_{\pm}$ and/or μ.

6.8   Lévy Stable Distributions in Real Space

Imagine that the legend of Fig. 6.7 is removed. How would you be able to match each curve to its corresponding value of β and μ? *Hint:* how does the value of μ affect the "spikeness" of the distribution in real-space?

6.9   Tails of Lévy Stable Distributions

(a)   What is the form of the *tails* of a Lévy stable distribution corresponding to some β, μ?

(b)   What happens to the tail when we take a sum of n variables from such a distribution?

6.10   Constraints on Lévy Stable Distributions

The constraints of the RG approach give a potential characteristic function of the form (for $\beta = 0$) $e^{-C|\omega|^{\mu}}$.

Show that values $\mu > 2$ do not lead to a probability distribution (i.e., that there is no probability distribution the characteristic function of which takes the above form with $\mu > 2$). *Hint:* evaluate $\langle x^2 \rangle$.

Problem credit: Prof. Eli Barkai.

6.11   Lévy Stable Noise

Consider the following discrete realization of a Langevin equation:

$$x_{n+1} = x_n + f(x_n)\Delta t + \Delta t^{\alpha} \xi,$$

where ξ is a random variable drawn from a Lévy stable distribution with $\beta = 0$ and heavy-tail corresponding to $\mu < 2$. Here the forcing $f(x)$ will be either linear in x (corresponding to a quadratic confining potential) or cubic in x.

(a)   What should α be so that we may be able to change the value of Δt (e.g., take $\Delta t \to 0$) but nevertheless have a well-defined (consistent) statistics of $x(t)$?

(b)   For a *linear* restoring force $f(x) = -kx$, find the stationary distribution.

(c)   Simulate the process numerically for $\mu = 1$ and a *cubic* restoring force $f(x) = -x^3$ for times $0 < t < 10^4$ to estimate the stationary distribution. Use a time step $\Delta t = 0.0001$. Note that for a given Δt, the discrete Langevin equation approximates the continuous one only as long as $\Delta t \cdot x^3$ is small (since the dynamics must be smooth). For $\Delta t = 10^{-4}$, this implies a maximal x of $10^{4/3} \approx 20$. Once the particle gets too far away we should discard the rest of the trajectory. When the distance from the origin exceeds 10 for the first time, you should estimate the stationary distribution from the trajectory up to that time. At the next time step, reset the particle position by choosing a position in $[-10, 10]$ according to your estimate. Repeat this process each time the distance from the origin exceeds 10.

Is the result you got surprising compared with the stationary distribution for this potential for Gaussian noise? If so, why?

Hint: to generate Cauchy-distributed variables you can use a uniformly distributed variable in the interval $(-1/2, 1/2)$ and then use the transformation $y = tan(\pi x)$.

Note: this problem is motivated by A. V. Chechkin, J. Klafter, V. Y. Gonchar, R. Metzler, and L. V. Tanatarov, Bifurcation, bimodality, and finite variance in confined Lévy flights. *Physical Review E*, 67(1), p.010102 (2003).

6.12   Record Statistics

At every time step, a variable x is drawn from the distribution $p(x)$.

(a)   What is the probability that at time step N a new "record" forms, i.e., the variable drawn is larger than any of the ones drawn beforehand?

(b)   What is the expected number of records until some time $N \gg 1$?

6.13 Scaling for Gaussian

(a) The error function erf(x) is defined as erf(x) $= 1 - \frac{2}{\sqrt{\pi}} \int_x^\infty e^{-t^2} dt$. Use integration by parts to find an explicit approximation for the error function using Gaussians, valid for large x. Use this to write an approximate expression for the cumulative distribution $G_n(x)$ for the maximum of n i.i.d. standard normal variables.

(b) The maximum of n normal variables will converge to a Gumbel distribution after appropriate scaling, i.e., $G_n(x) \to G(\frac{x-b_n}{a_n})$ as $n \to \infty$, where $G(x) = e^{-e^{-x}}$ is the Gumbel cumulative distribution. Based on your results in (a), derive a set of equations involving a_n and b_n from the requirement that $\lim_{n\to\infty} G_n(a_n x + b_n) = G(x)$.

(c) Show that the following choice for the scaling coefficients,

$$a_n = 1/b_n; b_n = \sqrt{2\log(n) - \log(4\pi \log(n))}$$

is an adequate choice that solves the equations derived in (b). Is the choice unique?

6.14 Black Swans

Consider n i.i.d. positive random variables $X_1, \ldots, X_n$ (where n is not necessarily large) whose distribution has a power-law tail $p(x) \propto x^{-(1+\mu)}$ with $0 < \mu < 2$.

(a) Find the tail of the distribution of the sum S_n of the n variables to leading order.

(b) Find the tail of the distribution of the maximum M_n of the n variables to leading order.

(c) Conditioned on the event that the sum S_n is very large, what is the probability that this is due to a single summand being very large, of the order of the entire sum – in other words, due to a single big jump?

(d) Verify the result in (c) numerically.

Problem credit: Pétur Rafn Bryde.

6.15 No Convergence to a Limiting EVD

Consider a probability distribution whose support is (e, ∞) and with the cumulative: $C(x) = 1 - 1/\log(x)$. Show that no linear scaling can lead to convergence to *any* of the three universal limiting distributions.

7 Anomalous Diffusion

> Now, here, you see, it takes all the running you can do, to keep in the same place. If you want to get somewhere else, you must run at least twice as fast as that!
>
> (Lewis Carroll, *Alice Through the Looking Glass*)

A fascinating application of the generalized central limit theorem and the Tauberian theorems appears in the context of *anomalous diffusion*. These are situations where a particle performs a random walk, but either the mean time between jumps is infinite, or the variance of the step size is – in both cases the assumptions we made early on in the book when deriving the diffusion equation (following Einstein's seminal work) are invalid – and indeed, these will lead to "subdiffusion" and "superdiffusion," respectively (i.e., $x^2 \sim t^\alpha$, with $\alpha \neq 1$).

The concept of anomalous diffusion was first introduced in the context of the motion of particles (in fact, holes) in semiconductors: In a seminal work, Harvey Scher and Elliott W. Montroll (1975) were studying the dynamics of the hole motion at Xerox, since it was relevant for the physics of photocopy machines. They realized that the experimental measurements were inconsistent with the standard scaling $x^2 \sim t$, but is perfectly consistent with a model where $x^2 \sim t^\alpha$ with $\alpha < 1$. This led them to consider a model where there are "traps" in the material, where in order to escape the particle has to overcome a barrier with energy U, and they assumed an exponential distribution of U, i.e.,

$$P(U) \propto e^{-U/U_0}. \tag{7.1}$$

Since the rate of escape is exponential in U, the associated timescale is

$$t \propto e^{U/kT}, \tag{7.2}$$

implying that the distribution of trapping times $\psi(t)$ is given by

$$\psi(t) = e^{-U/U_0} / \left| \frac{dt}{dU} \right| \propto e^{-U/U_0}/t = t^{-\frac{kT}{U_0}-1}. \tag{7.3}$$

This is a power-law distribution of trapping times, which potentially may have an infinite mean (when $kT \leq U_0$). Note that we just saw one of the "simplest" ways of rationalizing a power-law distribution: The random variable depends exponentially on another random variable that is exponentially distributed.

7.1   Continuous Time Random Walks

We shall now explore the implications of such a scenario, assuming, for simplicity, that the random walker is in 1D and makes a step of constant magnitude to the right or to the left at every step. We will further assume that at every site the particle will be trapped for a time t drawn from a distribution with a power-law tail $\psi(t) \propto 1/t^{1+\mu}$. Note that the fact that time is now a *continuous* variable is fundamentally different from the discrete time approach we took in Chapter 2 (in the context of Einstein's derivation of the diffusion equation). This formalism is known as the *continuous time random walk* (CTRW), and popularized by the seminal work of Scher and Montroll (1975). Our derivation will be similar to that of the excellent textbook focusing on anomalous diffusion (Klafter and Sokolov 2011).

Notation, notation ... Note that in contrast to Chapter 6, here we shall denote the Laplace transform of $\psi(t)$ by the *same* function, albeit with a different argument (s instead of t). The reason for this is that shortly we will consider the Laplace transform of several other distributions, and so using the same notation for both the distribution *and* its Laplace transform will save us from excessive notation. Whenever the argument of the function is s, you may assume it is the Laplace transform. This notation also follows closely that of Klafter and Sokolov (2011).

Consider the Laplace transform $\psi(s)$ of $\psi(t)$. According to what we showed in Section 6.4, if the mean of the distribution exists, then for small s,

$$\psi(s) \approx 1 - s\langle t \rangle, \tag{7.4}$$

while if the tail is heavy such that the mean diverges ($\mu \leq 1$), we have

$$\psi(s) \approx 1 - as^{\mu}, \tag{7.5}$$

where the probability distribution has a power-law tail proportional to $1/x^{1+\mu}$.

Consider the *survival probability* $\Psi(t)$, which we will define as the probability not to have moved at all by time t.

$$\Psi(t) = \int_t^{\infty} \psi(t')dt' = 1 - \int_0^t \psi(t')dt'. \tag{7.6}$$

Taking the Laplace transform, we find that

$$\Psi(s) = \frac{1}{s} - \frac{\psi(s)}{s} = \frac{1 - \psi(s)}{s}. \tag{7.7}$$

Next, let us consider the probability distribution for the time of the nth step, $\psi_n(t)$. It is straightforward to see that

$$\psi_n(t) = \int_0^t \psi_{n-1}(t')\psi(t - t')dt'. \tag{7.8}$$

Taking the Laplace transform of this convolution, we find that

$$\psi_n(s) = \psi_{n-1}(s)\psi(s), \tag{7.9}$$

and by iterating this n times we find that

$$\psi_n(s) = [\psi(s)]^n. \tag{7.10}$$

If instead we consider the probability to take n steps any time *up to* time t, which we will denote by $x_n(t)$, we have

$$x_n(t) = \int_0^t \psi_n(t')\Psi(t - t')dt', \tag{7.11}$$

and since this is also a convolution, we find that

$$x_n(s) = \psi_n(s)\Psi(s) = \psi^n(s)\frac{1 - \psi(s)}{s}. \tag{7.12}$$

Thus, we expressed this quantity only in terms of the Laplace transform of $\psi(t)$.

Now we are in a position to calculate the average number of steps taken until time t:

$$n(t) = \sum_{n=1}^{\infty} n x_n(t). \tag{7.13}$$

Taking the Laplace transform and plugging in the formula for $x_n(s)$ we find that

$$n(s) = \sum_{n=1}^{\infty} n\psi^n(s)\frac{1 - \psi(s)}{s} = \frac{1 - \psi(s)}{s}\sum_{n=1}^{\infty} n\psi^n(s). \tag{7.14}$$

We have seen this sum in Chapter 1, and it is easy to sum by writing it as the derivative of a geometric series:

$$n(s) = \frac{1 - \psi(s)}{s}\left[\psi(s)\frac{d}{d\psi(s)}\sum_{n=0}^{\infty} \psi^n(s)\right] = \frac{1 - \psi(s)}{s}\psi(s)\frac{1}{(1 - \psi(s))^2}. \tag{7.15}$$

Therefore,

$$n(s) = \frac{\psi(s)}{s}\frac{1}{1 - \psi(s)}. \tag{7.16}$$

For small s, and $\mu < 1$, we have seen that $\psi(s) \approx 1 - as^\mu$, and we obtain

$$n(s) \propto \frac{1}{s^{1+\mu}}, \tag{7.17}$$

which in turn tells us that for sufficient long times:

$$n(t) \propto t^\mu, \tag{7.18}$$

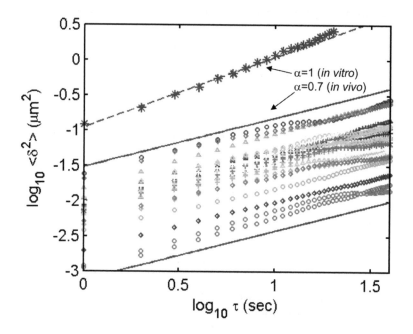

Figure 7.1 Subdiffusion observed in a living cell. Reprinted figure with permission from Golding and Cox (2006). Copyright 2006 by the American Physical Society.

with $\mu < 1$, since the Laplace transform of a power-law t^μ is proportional to $\frac{1}{s^{1+\mu}}$ (this is essentially another Tauberian theorem – try to prove it more systematically!). In fact, we could have guessed this scaling since after n steps, the distribution of the total time, T_n, would be the sum of n random variables drawn from the distribution $\psi(t)$, and hence according to the generalized CLT the scaled variable

$$\xi \equiv T_n/n^{1/\mu} \tag{7.19}$$

converges to a Lévy stable distribution (i.e., it is *independent* of n).

If the step size distribution has a finite variance, this would lead to subdiffusive behavior, where

$$\langle x^2 \rangle = n(t) \propto t^\mu. \tag{7.20}$$

Such behavior is observed in numerous cases, well beyond the semiconductors where the original discovery was made. In particular, over the last decade they have been observed often in biological cases, such as the (anomalous) diffusion of proteins in the crowded environment inside the cell; see Fig. 7.1 for one such example.

Subordination and operational time Note that in our model time and position are decoupled, in the sense that if we know that we took n steps the problem of finding the probability distribution of the walker's position is obviously reduced to the one we solved in Chapter 2 (though the *time* to take these n steps is different). In other words, if we had a clock which "ticks" every time a move is made, then we will get normal diffusion. Such a clock is said to run on *operational* time. Clearly, if we know the relation between the operational time and the position distribution, as well as the relation of the operational and actual time, we have the information needed to find the temporal dynamics, which is exactly what we did above. Random walks that are decoupled in this way are referred to as being subordinated (Klafter and Sokolov 2011).

You may be concerned that the model studied above, where the trapping time is chosen *randomly* at each step, may not reflect the physical reality where the depth of the traps is not time-dependent. Amir, Oreg, and Imry (2010) shows that also in this so-called quenched disorder scenario subdiffusion arises, albeit via a rather different mathematical approach.

7.2   Lévy Flights: When the Variance Diverges

The opposite case also exists, where the diffusion is faster than normal. Consider a random walker in 1D, with a step size distribution described by $p(\Delta)$, which potentially has infinite variance, and we will assume to be symmetric (i.e., it has vanishing mean). After the walker has taken n steps, its position will be a sum of n random variables, each drawn from the distribution $p(\Delta)$.

According to the standard CLT, if the distribution has a finite variance, then the sum converges to a Gaussian with variance linear in n – and we obtain normal diffusion. However, if the distribution has a sufficiently heavy power-law tail and the variance diverges, then the sum will converge to a Lévy stable distribution. In this case we have seen that the scaled variable

$$\xi \equiv \sum x_i / n^{1/\mu} \tag{7.21}$$

converges to a Lévy stable distribution. Note that since the step size distribution is assumed to be symmetric, we didn't have to shift the sum in order to center it around zero. Therefore, the distance scales as $n^{1/\mu}$, with $\mu < 2$, implying that the diffusion is faster than normal. Fig. 6.1(c) shows a simulation of a particular run for $\mu = 1/2$, showing why this is called a Lévy flight – the sum is dominated by rare events. We can understand this based on the results of Chapter 6, where we showed that the maximum of N variables also scales as $n^{1/\mu}$ – the very same scaling as the sum in its entirety! This idea is nicely demonstrated in Fig. 7.2, showing a Lévy flight in two dimensions.

Lévy walks In the Lévy flights discussed above, the jump itself takes no time at all. A variant of this problem is known as a Lévy *walk*, where the finite duration of the large jump is taken into account (e.g., by assigning the walker a constant velocity). This difference is also discussed in Problem 7.2.

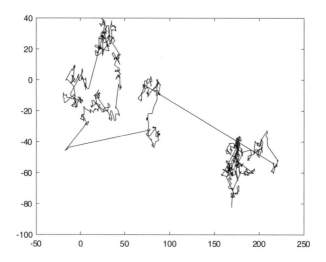

Figure 7.2 A two-dimensional Lévy flight, where each step size is drawn from a heavy-tailed distribution (with $\mu = 1.5$), and with a randomly and uniformly chosen angle.

7.3   Propagator for Anomalous Diffusion

In fact, we can get more than the scaling behavior – let us now derive the propagator $p(x,t)$, as we have done for the case of normal diffusion (where we found it to be a Gaussian).

$$p(x,t) = \sum_{n=0}^{\infty} p_n(x)x_n(t), \qquad (7.22)$$

since $x_n(t)$ denotes the probability of taking n steps until time t, and $p_n(x)$ is the probability distribution of position if we took n steps. Taking the Laplace transform with respect to time gives us

$$p(x,s) = \sum_n p_n(x)x_n(s) = \sum_n p_n(x)\psi^n(s)\frac{1 - \psi(s)}{s} = \frac{1 - \psi(s)}{s}\sum_n p_n(x)\psi^n(s).$$

$$(7.23)$$

Taking the Fourier transform with respect to space gives us

$$p(k,s) = \frac{1 - \psi(s)}{s}\sum_n p^n(k)\psi^n(s), \qquad (7.24)$$

where $p(k)$ is the Fourier transform of the step size distribution $p(\Delta)$. Summing the geometric series gives us the propagator in Laplace/Fourier space:

$$p(k,s) = \frac{1 - \psi(s)}{s}\frac{1}{1 - p(k)\psi(s)}. \qquad (7.25)$$

7.4 Back to Normal Diffusion

For example, if the time distribution has a mean and the step size distribution has a variance, we have

$$\psi(s) \approx 1 - \langle t \rangle s; \ p(k) \approx 1 - \frac{\langle x^2 \rangle}{2} k^2. \tag{7.26}$$

Plugging these into Eq. (7.25) gives

$$p(k,s) \approx \frac{1}{s + \frac{\langle x^2 \rangle}{2\langle t \rangle} k^2}. \tag{7.27}$$

Taking the inverse Laplace gives

$$p(k,t) = e^{-\frac{\langle x^2 \rangle}{2\langle t \rangle} k^2 t}, \tag{7.28}$$

and taking the inverse Fourier gives

$$p(x,t) \propto e^{-\frac{\langle t \rangle}{2\langle x^2 \rangle} \frac{x^2}{t}}, \tag{7.29}$$

which is precisely the solution to the diffusion equation with

$$D = \frac{\langle x^2 \rangle}{2\langle t \rangle}. \tag{7.30}$$

This generalizes the derivation of the diffusion equation which we presented in Chapter 2, to cases where both the step size and the duration of each step are random variables – we see that as long as the distributions are not heavy-tailed, we get normal diffusion.

7.5 Ergodicity Breaking: When the Time Average and the Ensemble Average Give Different Results

Let us go back to the scenario where the step size is "normal," but the mean trapping time diverges. We will now describe an intriguing phenomenon following He *et al.* (2008), where ensemble averaging and time averaging lead to a different result in the case of anomalous diffusion.

Within the CTRW model discussed previously, in one dimension, consider tracking a single particle over time, where the trapping time distribution has infinite mean. If we look at the mean square displacement over a time window Δ, starting at time 0 and ending at time T, we have

$$\langle\langle x^2 \rangle\rangle_{time} = \frac{1}{T - \Delta} \int_0^{T-\Delta} \langle [x(z + \Delta) - x(z)]^2 \rangle dz. \tag{7.31}$$

(Note that we have assumed that the time-averaging procedure is repeated over many experiments, leading to the additional ensemble-averaging denoted by $\langle \rangle$.)

Previously, we saw that the simple procedure of *ensemble average* (i.e., the result of repeating the experiment many times and averaging over different runs) gives

$$\langle\langle x^2\rangle\rangle_{ensemble} \propto \langle n(t)\rangle \propto t^\mu. \tag{7.32}$$

$n(t)$ was the expected number of steps made until time t (as noted this is also known as the "operational time" – since it is the only quantity which matters with regards to the position distribution).

Since the mean $\langle x(t)\rangle$ vanishes at any time, we have

$$\langle[x(t_2) - x(t_1)]^2\rangle \propto \langle n(t_2)\rangle - \langle n(t_1)\rangle \tag{7.33}$$

(i.e., the additional number of steps made between these two times is the quantity which matters and leads to the difference in position between these two measurements.) Plugging this into the integral of Eq. (7.31) we find that

$$\langle\langle x^2\rangle\rangle_{time} \propto \frac{1}{T - \Delta} \int_0^{T-\Delta} [(z + \Delta)^\mu - z^\mu]dz. \tag{7.34}$$

Evaluating the integral we find that

$$\langle\langle x^2\rangle\rangle_{time} \propto \frac{1}{T - \Delta}[T^{1+\mu} - \Delta^{1+\mu} - (T - \Delta)^{1+\mu}]. \tag{7.35}$$

In the limit $\Delta \ll t$ (we track a single particle over a long time), we find that

$$\langle\langle x^2\rangle\rangle_{time} \propto \frac{1}{T^{1-\mu}}\Delta. \tag{7.36}$$

This implies we will measure *normal* diffusion when doing a time average (rather than an ensemble average), but the diffusion constant will depend on the time over which we track the particle, becoming smaller and smaller as we measure over larger time windows! This phenomenon is explored numerically in Problem 7.4.

Another striking manifestation of heavy-tailed distributions is explored in Problem 7.5: The so-called inspection paradox shows a discrepancy between lightbulbs having a finite-lifetime, albeit when we *observe* installed lightbulbs the mean lifetime appears to be arbitrarily large . . .

7.6   Summary

In Chapter 2, we derived the diffusion equation following Einstein's approach, where we made two assumptions: The step size distribution has a finite variance, and the time step is constant. In this chapter we relaxed both of these assumptions, and found that they lead to rather intriguing differences with the diffusive behavior arising from the former case. We introduced the continuous time random walk formalism, and saw that if the time step distribution is sufficiently heavy-tailed we obtain subdiffusion. Interestingly, in this case the ensemble average and the time average are distinct (a phenomenon known as ergodicity breaking). In the case of a stepsize with infinite variance, we saw that the distribution is no longer Gaussian but rather converges to a Lévy stable distribution. Moreover, in this case the random walker's position is dominated by rare events – this is known as a Lévy flight. All of these "exotic" behaviors of random walks have ample realizations in natural phenomena, reinforcing a quote oft used by physicists "Everything not forbidden is compulsory."

For further reading Klafter and Sokolov (2011) provides a more detailed and very accessible discussion of the topics covered in this chapter. Hughes (1996) provides a more mathematical discussion of the topics discussed here (such as the Scher–Montroll model), including other models for non-diffusive behavior not touched upon here.

7.7   Exercises

7.1   Random Walking in Combville

The (infinite) city Combville consists of an infinite main road, along which, at regular intervals of size 1, are infinite side walks (see Fig. 7.3).

A walker performs a random walk in the city, where at every time step if they are on one of the junctions on the main road, they take a unit step with equal probability to each of the three possible directions, and otherwise they take a step up or down along the side street with equal probability (e.g., a possible route is $(0,0) \rightarrow (1,0) \rightarrow (1,1) \rightarrow (1,2) \rightarrow (1,1) \rightarrow (1,0) \rightarrow (2,0)\ldots$).

(a) If a walker just entered a side street, what is the probability distribution function for returning to the main street? How long would it take them to return on average?

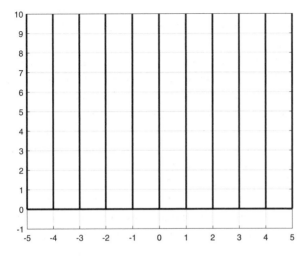

Figure 7.3 Roads in Combville.

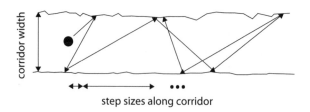

Figure 7.4 Schematic of particle motion in narrow corridor.

(b) If the walker started on the main road at $x = 0$, how would their mean squared position along the main road scale with time?

7.2 Lévy Walks

(a) Consider a gas particle moving in an infinite 2D corridor which moves straight at a constant velocity and bounces off the walls at a random angle drawn from a uniform distribution, as may be expected to happen if the corridor walls were corrugated. What is the probability distribution $p(\Delta)$ of step sizes Δ along the corridor (see Fig. 7.4 for clarification)?

(b) Find the PDF of the position x after N steps. In the context of the Generalized Central Limit Theorem, what is the appropriate scaling of x?

(c) Assuming that the particle moves at a constant speed v, find the PDF of the time elapsed per step, τ. If the number of steps N is large and finite, what is the appropriate scaling and the limiting behavior of the PDF of the total time t?

This problem was written by Felix Wong.

7.3 Adaptive Walk on Fitness Landscapes*

The adaptation of an organism to its environment arises in part by alleles having different effects on the organism's "fitness," or ability to reproduce. Assuming that we can assign a value (a "fitness effect") to quantify the effect of a beneficial genetic mutation, what does the distribution of these fitness effects look like? Extreme value theory provides a clue.

Consider m possible single mutations that may occur in a wild-type population of organisms. Thus, there are a total of $m + 1$ possible phenotypes, including the unmutated wild-type (which can mutate into any of the other m). Let us order all $m + 1$ phenotypes along a "fitness line," where the fittest allele has rank 1 and the gap between phenotypes $i, i + 1$ is denoted G_i. A beneficial mutation improves the fitness by an amount ΔW, which can be expressed as a sum of corresponding G_i's. In particular, we define the fitness advantage of an allele k relative to the allele i (with $i > k$) as the sum of the relevant spacings $\Delta W_k = G_{i-1} + G_{i-2} + \cdots + G_k$.

Since the wild-type has dealt with evolutionary pressure well up until the mutation event, we also assume that most of the m mutations will be less fit than the wild-type. The adaptation we consider is therefore confined only to the tail of the distribution of fitnesses. Suppose that we do not know the values of the G_i, but only that the distribution of fitness values is drawn from a Gaussian distribution.

(a) If the distribution of fitness values, $p(x)$, is Gaussian, what is the asymptotic distribution of fitness advantages among beneficial mutations, ΔW_k, for all $k < i$? In other words, consider drawing N random variables from a Gaussian distribution, look only at the M "fittest" ones, with $N \gg M$, and determine, analytically, the asymptotic distribution of the fitness advantages.

(b) Show that this result holds for any $p(x)$ in the Gumbel universality class.

This problem is based on H. A. Orr, The distribution of fitness effects among beneficial mutations. *Genetics*, 163(4), pp. 1519–1526 (2003).

7.4 Numerically Observing Ergodicity Breaking

(a) Simulate a 1D random walker where every step is ±1 with equal probability, and the time interval between steps is drawn from the distribution $p(t) \propto 1/t^{1+\mu}$ for $t > 1$, with $\mu < 1$. Calculate numerically and compare the ensemble average $\langle\langle x^2(t)\rangle\rangle_{\text{ens}}$ and the time average

$$\langle\langle x^2 \rangle\rangle_{\text{time}} = \frac{1}{T - \Delta} \int_0^{T-\Delta} \left[x(z + \Delta) - x(z)\right]^2 dz.$$

For the ensemble average, plot $\langle\langle x^2(t)\rangle\rangle_{\text{ens}}$ as a function of time for $0 < t < 10^4$. For the time average, generate a single trajectory $x(t)$ with length 10^9 and plot $\langle\langle x^2 \rangle\rangle_{\text{time}}$ as a function of the time window Δ, for Δ between 10^3 and 10^5.

(b) Now simulate a random walk where the time interval between steps is fixed, but the step size is drawn from a Cauchy distribution. Estimate the following quantities by averaging over 10^4 runs and plot for $0 < t < 10^4$: (i) $\langle\langle\sqrt{|x|}\rangle\rangle^2$, (ii) $\sqrt{\langle\langle x^2\rangle\rangle}$. Explain the difference in behavior you observe.

7.5 Rare Events and the Inspection Paradox

A factory makes lightbulbs, whose lifetime is a random variable drawn from the distribution $p(t) \propto 1/t^{1+\mu}$ for $t > 1 day$, with $1 < \mu < 2$.

Bob uses lightbulbs from the factory in his house, replacing them every time they burn out.

(a) How much time, approximately, would it take Bob to go through 100 bulbs?

(b) We pay Bob a visit at a random time, chosen uniformly in an interval $[0, T]$ (with $T \gg 1 day$). What will be the mean lifetime of the lightbulb currently installed in Bob's house?

Note that this problem is a more dramatic manifestation of the sort of effect we observed in Problem 1.4 of Chapter 1! This is called "the inspection paradox," for reasons that should be clear once you solve this problem.

7.6 Facilitated Diffusion*

Consider a protein bound to a planar surface, suspended in a liquid. Every once in a while the protein unbinds from the surface, and does a (normal) 3D diffusion in the liquid. If it hits the planar surface again, it will bound to it until the next unbinding event.

(a) What is the probability for the protein to return to the surface? What is the distribution of return times?

(b) Consider the series of binding points on the two-dimensional surface. What is the distribution $p(\vec{r})$ that describes the *two-dimensional* step taken from one binding position to the next?

(c) What is the position distribution on the 2D surface after $N \gg 1$ unbinding–binding events have taken place?

8 Random Matrix Theory

The miracle of the appropriateness of the language of mathematics for the formulation of the laws of physics is a wonderful gift which we neither understand nor deserve.

(Eugene Paul Wigner, 1960)

8.1 Level Repulsion between Eigenvalues: The Birth of RMT

In the 1950s, physicists measured the nuclear energy levels of complex atoms, and found to their surprise that the probability distribution of the spacing between two neighboring energy levels is far from Poissonian – in fact, the probability distribution appeared to vanish at zero level spacing, a phenomenon known as "level repulsion." This is shown in Fig. 8.1. In quantum mechanics, the energy levels correspond to the eigenvalues of the Hamiltonian, but deriving them for these complex nuclei is out of reach. How can we understand the level repulsion without obtaining a detailed, system-specific solution? In 1955 Eugene Wigner had an ingenious solution to circumvent this problem, which led to the flourishing field of random matrix theory (RMT). His insight was that to model certain aspects of such complex systems, we may consider the Hamiltonian as a *random* entity, i.e., a matrix whose entries will be independent random variables (yet respecting the symmetries of the problem – the Hamiltonian has to be a Hermitian operator). In this chapter we shall see how this simple picture leads to a specific form of level repulsion that captures the experimental data very well, and was later found to describe numerous other scenarios in physics and mathematics. We will also calculate the distribution of eigenvalues of such matrix ensembles, and find the joint probability distribution of the eigenvalues. Finally, we will extend some of the results to *non-Hermitian* random matrices, with applications to models in ecology and neuroscience. A recent book provides an excellent extended (yet very approachable) discussion of the first part of this chapter (on Hermitian random matrices) (Livan *et al.* 2018).

8.1.1 Defining the Random Matrix Ensembles

Consider the set of real, symmetric matrices, with a probability density scaling as

$$p(\boldsymbol{H}) \propto e^{-\frac{N}{4}tr\boldsymbol{H}^2} = e^{-\frac{N}{4}\sum_{i,j}H_{ij}^2}. \tag{8.1}$$

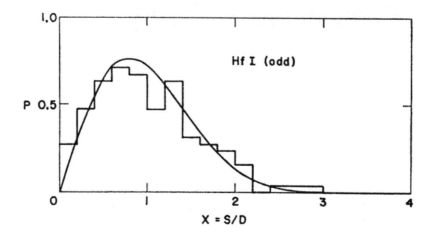

Figure 8.1 "Level repulsion" between energy levels of complex atoms, compared with the theoretical expectation due to Wigner's random matrix model. Reprinted Figure 3 with permission from Rosenzweig and Porter (1960). Copyright 1960 by the American Physical Society.

This ensemble is known as the Gaussian Orthogonal Ensemble (GOE). The use of the trace in the equation implies that changing to a new basis will not affect the probabilities, which is a rather natural property to expect. As we see from the RHS, this formulation is equivalent to taking each of the $(N^2 - N)/2$ matrix elements in the upper diagonal as independent, Gaussian variables with vanishing mean and variance $1/N$, while the diagonal elements are also independent, Gaussian variables with vanishing mean and variance $2/N$. Note that the $N/4$ prefactor is merely a normalization choice, which will determine the scale of the eigenvalues. Similarly, the Gaussian Unitary Ensemble (GUE) contains Hermitian matrices with complex entries, and the probability density is in this case

$$p(\boldsymbol{H}) \propto e^{-\frac{N}{2} tr H^2} = e^{-\frac{N}{2} \sum_{i,j} H_{ij} H_{ji}}. \tag{8.2}$$

Since the matrices are Hermitian, we have

$$\sum_{i,j} \boldsymbol{H}_{ij} \boldsymbol{H}_{ji} = \sum_{i,j} |\boldsymbol{H}_{ij}|^2, \tag{8.3}$$

hence this choice is equivalent to taking the real and imaginary parts of the matrix entries as independent Gaussian variables.

Other random matrix ensembles Throughout this chapter, both in the context of Hermitian and non-Hermitian random matrices, we will consider non-sparse matrices: i.e., the probability of vanishing matrix elements $\boldsymbol{H}_{ij}$ will be negligible. In many applications, however, the matrices are sparse or effectively sparse (with most matrix elements zero or exponentially small). One such class of random matrices arises in the context of "tight-binding" models in physics, where matrix elements are associated with the interactions between two sites in space, and the interaction falls off rapidly with the distance. Such models have been extensively studied in the context of Anderson localization (where the matrices correspond to the Hamiltonians of electrons in a disordered solid, see Anderson (1958)), phonons in disordered systems (e.g., Amir *et al.* 2013a), and many more, and exhibit rich phenomena distinct from those discussed here (e.g., how wave interference in disordered systems can, perhaps counterintuitively, destroys diffusion! (Anderson 1958)). Related non-Hermitian models have recently been studied in the context of neuroscience (Amir *et al.* 2016a) and other fields, see Metz *et al.* (2019) for a recent review. Finally, there are also many variants of the non-sparse ensembles studied in this chapter – see, for example, Problems 8.4 and 8.7 (for Hermitian and non-Hermitian matrix ensembles, respectively).

8.1.2   Wigner's Surmise – Simple Analytical Results for $N = 2$

Calculating the distribution of level spacing in the case $N \gg 1$ is rather lengthy and challenging. However, it turns out, surprisingly, that solving it for the case $N = 2$ provides an excellent approximation to the level spacing distribution for the case of large N, as Wigner insightfully noticed. This procedure is known as "Wigner's surmise." In this case, for the GOE our matrices will have three independent degrees of freedom, and have the structure

$$H = \begin{pmatrix} a & b \\ b & c \end{pmatrix} \tag{8.4}$$

with

$$p(a,b,c) \propto e^{-\frac{a^2+2b^2+c^2}{2}}. \tag{8.5}$$

Since we will be interested in the normalized level spacing, let's drop the factor of 2 in the denominator in what follows (which is equivalent to rescaling all matrix elements, and hence the eigenvalues, by $\sqrt{2}$). It is easy to find the normalization constant, leading to

$$p(a,b,c) = Ce^{-[a^2+2b^2+c^2]}, \tag{8.6}$$

with $C = \frac{\sqrt{2}}{\pi^{3/2}}$.

The eigenvalues of H are given by

$$\lambda_{1,2} = \frac{a+c}{2} \pm \frac{\sqrt{(a-c)^2 + 4b^2}}{2}. \tag{8.7}$$

Hence the level spacing is

$$\Delta\lambda = \sqrt{(a-c)^2 + 4b^2}, \tag{8.8}$$

and we need to calculate

$$p(\Delta\lambda) = \int_{-\infty}^{\infty}\int_{-\infty}^{\infty}\int_{-\infty}^{\infty} p(a,b,c)\delta(\Delta\lambda - \sqrt{(a-c)^2 + 4b^2})da\,db\,dc. \tag{8.9}$$

It is convenient to make the following change of variables:

$$
\begin{pmatrix} \tilde{a} \\ \tilde{b} \\ \tilde{c} \end{pmatrix} = \begin{pmatrix} 1 & 0 & -1 \\ 0 & 2 & 0 \\ 0 & 0 & 1 \end{pmatrix} \begin{pmatrix} a \\ b \\ c \end{pmatrix}.
\tag{8.10}
$$

The Jacobian of the transformation is 1/2, and in terms of the new variables the integral we have to evaluate is

$$
p(\Delta\lambda) = \frac{C}{2} \int_{-\infty}^{\infty} \int_{-\infty}^{\infty} \int_{-\infty}^{\infty} e^{-[(\tilde{a}+\tilde{c})^2 + \tilde{c}^2 + 2(\tilde{b}/2)^2]} \delta\left(\Delta\lambda - \sqrt{\tilde{a}^2 + \tilde{b}^2}\right) d\tilde{a}\, d\tilde{b}\, d\tilde{c}.
\tag{8.11}
$$

We can readily perform the integration over $\tilde{c}$, recognizing that it is a Gaussian integral which can be evaluated by completing the square:

$$
\int_{-\infty}^{\infty} e^{-2\tilde{c}^2 - 2\tilde{a}\tilde{c}} d\tilde{c} = \int_{-\infty}^{\infty} e^{-2[\tilde{c}+\tilde{a}/2]^2} e^{\tilde{a}^2/2} d\tilde{c} = e^{\tilde{a}^2/2} \frac{\sqrt{\pi}}{\sqrt{2}}.
\tag{8.12}
$$

Plugging this back into Eq. (8.11), we find that

$$
p(\Delta\lambda) = \frac{C}{2} \frac{\sqrt{\pi}}{\sqrt{2}} \int_{-\infty}^{\infty} \int_{-\infty}^{\infty} e^{-\frac{\tilde{a}^2 + \tilde{b}^2}{2}} \delta\left(\Delta\lambda - \sqrt{\tilde{a}^2 + \tilde{b}^2}\right) d\tilde{a}\, d\tilde{b}.
\tag{8.13}
$$

Since this integral only depends on $\tilde{a}^2 + \tilde{b}^2$, it is very natural to make another change of variables – this time to polar coordinates. We then find that

$$
p(\Delta\lambda) = \frac{C}{2} \frac{\sqrt{\pi}}{\sqrt{2}} \int_{0}^{2\pi} \int_{0}^{\infty} r e^{-\frac{r^2}{2}} \delta(\Delta\lambda - r)\, dr\, d\theta.
\tag{8.14}
$$

Note the appearance of an extra r in front, which comes from the Jacobian of the transformation – this will lead to the most interesting aspect of the level spacing distribution, namely, the level repulsion (later we will see another case where the "action" occurs in the Jacobian of the transformation).

Evaluating this integral is straightforward, and plugging in the previous formula for C we find that

$$
p(\Delta\lambda) = \Delta\lambda\, e^{-\Delta\lambda^2/2}.
\tag{8.15}
$$

As a sanity check, you may check that this is a normalized probability distribution. We can see that the probability distribution vanishes linearly for small $\Delta\lambda$, which is the phenomenon of level repulsion. Finally, the average level spacing can be easily evaluated to be $\sqrt{\pi/2}$. When comparing this formula to, say, experimental data, we will want to scale the level spacing distribution to be dimensionless, by looking at the distribution of the level spacing scaled by the *average* level spacing. For this reason it is useful to rescale $\Delta\lambda$ of Eq. (8.15) to obtain a distribution with mean of 1:

$$
p(s) = \frac{\pi}{2} s\, e^{-\frac{\pi}{4} s^2}.
\tag{8.16}
$$

You will often see Wigner's surmise presented in this form.

Problem 8.2 compares the result of Eq. 8.16 to the level statistics of *large* matrices, finding remarkably good agreement between the two; $N = 2$ providing a good approximation to the $N \to \infty$ case! Note that in the case of $N = 2$, which we just solved, the fact that the variance of the diagonal elements was different from the off-diagonal ones was instrumental (and actually simplified the calculation). Interestingly, for large N this would not have made a difference – the same statistics will be obtained even if the diagonal elements are chosen with the same statistics as the off-diagonal ones, since their number pales in comparison to the $O(N^2)$ off-diagonal elements. Also note that for $N \to \infty$ the distribution of $\Delta\lambda$ can be derived *exactly* for the GOE, see Livan *et al.* (2018). The exact form is distinct from Wigner's surmise, but surprisingly close to it.

Where did it come from? One may get intuition for the "level repulsion" by direct inspection of Eq. (8.8). For a diagonal matrix, $b = 0$, and the levels will coincide when $a = c$. In the presence of the off-diagonal elements, *two* conditions need to be satisfied for the levels to coincide: in addition to $a = c$, b must vanish. This is what leads to the linear suppression of the distribution for small values of $\Delta\lambda$. Similarly, if the matrix elements were *complex*, both the real and imaginary part of the off-diagonal elements should vanish – hence the dependence for small $\Delta\lambda$ will be quadratic rather than linear. This is explored, analytically, in Problem 8.1.

Remarkably, many phenomena in physics, mathematics, and other disciplines follow closely the results of random matrix theory for the level repulsion. Several examples are illustrated in Fig. 8.2: (a) shows that the distribution of spacings between buses in the city of Cuernavaca in Mexico is well approximated by Wigner's surmise. This appears strange in light of the simple argument in Chapter 1 suggesting it should be an exponential distribution. It turns out that buses "interact" in this case, with an effective repulsion induced by people who inform bus drivers that another bus is slightly ahead of them and they should slow down. A few years after this fortuitous finding, it was proven that within a model of this interacting bus problem, the predictions of RMT rigorously hold! (Baik *et al.* 2006). (b) shows an example in pure mathematics, where the eigenmodes of the Laplacian operator are solved in a domain resembling a pool table (known as "Sinai's billiard"), and show level repulsion captured by the GOE random matrix ensemble.

8.2   Wigner's Semicircle Law for the Distribution of Eigenvalues

Wigner also explored the probability distribution for finding an eigenvalue at position λ ("empirical spectral distribution" in the mathematician's notation), which we will denote by $p(\lambda)$ (which may of course be a function of the size of the matrix N). It can

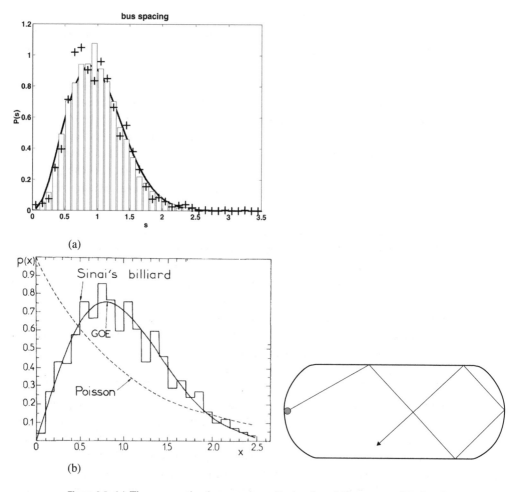

Figure 8.2 (a) Time separation between buses in the city of Cuernavaca, Mexico. From Krbálek and Seba (2000). © IOP Publishing. Reproduced with permission. All rights reserved. Surprisingly, the distribution is well captured by Wigner's surmise (for the GUE). (b) Level spacing distribution for the eigenvalues of the Laplacian operator in a "billiard-shaped" domain. Adapted from Bohigas *et al.* (1984). The chaotic nature of the classical trajectories (shown on the right) is related to the fact that the statistics here are well captured by the random matrix theory result.

be derived from the joint probability distribution of all N eigenvalues by marginal-izing over all but one of the eigenvalues. Even though we will later find an explicit expression for the joint probability distribution, here we will follow a different path (following roughly, in fact, the original approach due to Wigner (1955)). We will calculate (approximately) all moments of $p(\lambda)$ and show that they correspond to a "semicircle" – a probability distribution with finite support and with a $p(\lambda)$ tracing the shape of an arc. Fig. 8.3 shows the distribution of eigenvalues for a 3000×3000

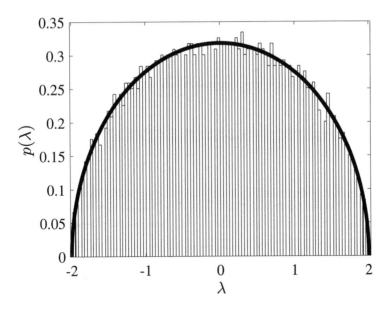

Figure 8.3 The distribution of eigenvalues for a single 3000×3000 matrix from the GOE ensemble, compared with the "semicircle law."

matrix chosen from the GOE. Obtaining such results numerically is extremely easy – in fact, it is quite typical in RMT that doing numerics is straightforward, yet analytical derivations are challenging.

Self-averaging Note that the simulation of Fig. 8.3 shows that the histogram of eigenvalues obtained from a diagonalization of a *single* large matrix is well approximated by the semicircle law. Often the mathematical derivations will involve *ensemble-averaging*, i.e., the results will pertain to the average over many realizations of the disorder. Yet in practice the result may be applicable to a single-realization of a large system – this property is often referred to as *self-averaging* (in other words, in this case the distribution of the DOS is highly concentrated around its ensemble-average, with small fluctuations).

In the following, we follow Wigner's derivation of the semicircle law, where the strategy is to calculate (approximately, for large N) the moments of the eigenvalue distribution we are seeking. Under quite general conditions (which are indeed applicable here) the moments uniquely specify the distribution, hence they allow us to recover the semicircle law. To proceed, consider the entity

$$M_k \equiv \left\langle \frac{1}{N} tr \, H^k \right\rangle, \tag{8.17}$$

where, as usual, $\langle\rangle$ denote ensemble-averaging. This can be written in terms of the eigenvalues of H as

$$M_k = \left\langle \frac{1}{N} \sum_{i=1}^{N} \lambda_i^k \right\rangle. \tag{8.18}$$

Expressing this in terms of the eigenvalue distribution, we find that

$$M_k = \int_{-\infty}^{\infty} p(\lambda)\lambda^k d\lambda, \tag{8.19}$$

where $p(\lambda)$ is the eigenvalue probability distribution (note that this relation holds also when the eigenvalues are correlated – which we know is actually the case from our discussion of level repulsion).

According to this equation, the kth moment of the eigenvalue distribution is proportional to the trace of H^k, and our trick will be now to calculate the latter in a different way. Let us first go through some examples. For $k = 1$, we have

$$\left\langle \frac{1}{N} tr\, H \right\rangle = 0, \tag{8.20}$$

since the matrix elements are drawn from a Gaussian distribution with vanishing mean. In fact, it is easy to see from symmetry that all odd moments of $p(\lambda)$ should vanish: If we have an eigenvalue λ with

$$H\vec{v} = \lambda\vec{v}, \tag{8.21}$$

then $-\lambda$ is an eigenvalue of $-H$ (with the same eigenvector), which has the same probability density in the GOE.

For $k = 2$, we have to consider

$$\left\langle \frac{1}{N} tr\, H^2 \right\rangle = \frac{1}{N} \left\langle \sum_{i,j} H_{ij} H_{ji} \right\rangle. \tag{8.22}$$

Since we have $\langle H_{ij} H_{ji} \rangle = 1/N$ for all $i \neq j$, we find that for large N (when the different variance of the diagonal elements will be insignificant),

$$M_2 \approx \frac{1}{N}\frac{1}{N}N^2 = 1. \tag{8.23}$$

Finally, let us consider the case of $k = 4$. A typical term in $tr\, H^4$ will be of the form: $H_{i_1 i_2} H_{i_2 i_3} H_{i_3 i_4} H_{i_4 i_1}$. However, in order for the ensemble average not to vanish, every term has to appear an even number of times. For example, we may have a structure where

$$i_2 \neq i_1; i_3 = i_1; i_4 \neq i_1, i_2. \tag{8.24}$$

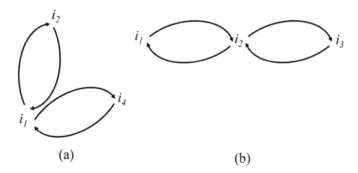

Figure 8.4 Two trees which contribute to the 4th moment.

This corresponds to the graph of Fig. 8.4(a), showing the distinct indices. In this case, there are three different distinct indices, and hence there are approximately N^3 options to choose them. Therefore, the contribution of all of these graphs to the 4th moment is

$$\frac{1}{N}N^3\frac{1}{N^2} = 1. \tag{8.25}$$

Similarly, the graph where $i_2 \neq i_1; i_3 \neq i_1, i_2; i_4 = i_2$ corresponds to the graph of Fig. 8.4(b), and by the same reasoning will also have a contribution of 1 to the moment. Finally, consider cases such as

$$i_2 \neq i_1; i_3 = i_1; i_4 = i_2, \tag{8.26}$$

i.e., cases where the number of distinct indices is *smaller*. In this case the contribution will be $O(1/N)$ or lower, and hence they will be negligible. You may similarly neglect scenarios in which the diagonal elements appear ("self-loops"). We therefore only have to consider cases where each matrix element appears twice, and we have the maximal number of distinct indices, namely, $k/2 + 1$. If we draw a directed graph starting from the first index (the "root" of the graph) and proceed by the order specified by the product of matrix elements (i.e., $H_{5,6}$ implies going from vertex 5 to 6), the aforementioned conditions imply that the directed graph always has an edge $i \rightarrow j$ simultaneously occurring with an edge $j \rightarrow i$. Furthermore, if we create a new *undirected* graph where each of these pairs of edges maps to a single edge, it must be a tree (i.e., contain no loops) since otherwise we would not be able to "explore" the maximal number of vertices (try to prove this rigorously, e.g., by induction). If – as in the above examples – we "pool together" all of the graphs which share the same topology (structure), the contribution to the moment will then be

$$\frac{1}{N}\left(\frac{1}{N}\right)^{k/2} N^{k/2+1} = 1. \tag{8.27}$$

Note that in Fig. 8.4, the graphs corresponding to (a) and (b) both look the same (they are isomorphic). Yet it is clear that they are "shorthand notation" for two distinct sets of graphs: For instance, $H_{3,4}H_{4,7}H_{7,4}H_{4,3}$ belongs to (b), but $H_{3,4}H_{4,3}H_{3,7}H_{7,3}$

belongs to (a). We may nevertheless characterize each topology uniquely by a *rooted* tree, where indeed in Fig. 8.4 the root is the middle vertex for (b) but not for (a). We conclude that the kth moment will be approximately an integer (this will become exact as $N \to \infty$), with a value equal to the number of distinct trees of this form.

To proceed, we will show a mapping between the set of these trees and *Dyck paths*, which are sequences of ± 1 whose running sum is always nonnegative, and with equal number of $+1$ and -1. Consider a sequence of the sort:

$$H_{i_1 i_2} H_{i_2 i_3} H_{i_3 i_4} H_{i_4 i_1}. \tag{8.28}$$

Every time we see a pair of indices i, j (without considering their ordering within this pair), it is either the first time we see it or the second time we see it – since each edge in the directed graph appears precisely once but its "partner edge" (with indices flipped) must also appear. We can associate the first event with $+1$, and the second with -1. Clearly, the running sum must be nonnegative. This maps a tree (i.e., sequence) to a Dyck path. The opposite mapping is also possible: Based on the sequence of ± 1 we can easily reconstruct the sequence of indices and its corresponding tree. Hence the mapping is one-to-one, and the problem boils down to finding the number of Dyck paths of length $2R$, containing an equal number of ± 1 (where here $2R = k$, the index of the moment we are interested in calculating).

This result is related to "Bertrand's ballot problem": Consider an election where candidate A receives a votes and candidate B receives b votes with $a > b$, what is the probability that A will be strictly ahead of B throughout the count of the votes? The answer is given by the ratio

$$p = \frac{a - b}{a + b}. \tag{8.29}$$

There are several ways to derive this result. An elegant method to enumerate the sequences uses the so-called reflection trick (Addario-Berry and Reed 2008). The first part of Problem 2.1 of Chapter 2 provides guidance to another method to do the accounting (see also Kostinski and Amir 2016).

How can we relate this result to our problem? If we add one more "1" before our sequence, then clearly every "legitimate" sequence (i.e., Dyck path) will become strictly positive (not just nonnegative) and every positive sequence of the new version corresponds to a nonnegative sequence in the previous one. Since the total number of paths is $\binom{a+b}{a}$, the number of "legitimate" sequences is given by

$$\binom{a + b}{a} \frac{a - b}{a + b}. \tag{8.30}$$

The number of Dyck paths is given by Eq. (8.30) with $a = 1 + R, b = R$:

$$I_R = \binom{2R + 1}{R} \frac{1}{2R + 1} = \binom{2R}{R} \frac{1}{R + 1}. \tag{8.31}$$

The numbers in this sequence are known as the *Catalan numbers* and appear in a wealth of applications in mathematics and computer science. According to what we have shown earlier we have $I_R = M_{2R}$. Problem 8.3 shows that these numbers are precisely the even moments of the "semicircle" law, i.e., the probability distribution:

$$p(x) = \frac{1}{2\pi}\sqrt{4 - x^2}. \tag{8.32}$$

Why $\sqrt{N}$? If the off-diagonal elements of the Hermitian matrix have a variance of 1 (rather than $1/N$), the radius of the semicircle law would be scaled by an additional $\sqrt{N}$ factor, to become $2\sqrt{N}$. We can get some intuition for the scaling by considering an eigenvector of this matrix. Being naive, we may expect that the product of a row of the random matrix by this vector would correspond to a random sum of numbers, so would scale as $\sqrt{N}$ multiplied by the typical value of an entry of the eigenvector – which elucidates the scaling of the eigenvalues with N (the additional factor of 2 goes beyond this simple scaling argument). The same logic would naturally also apply to the non-Hermitian case, and we will indeed find that the $\sqrt{N}$ scaling is also applicable there – albeit in the complex plane.

On the notion of universality We have (non-rigorously) derived the semicircle law, under the assumption that the matrix elements are drawn from the GOE (in particular, they are normally distributed). In fact, the law can be proven mathematically under much less stringent assumptions, and for instance the matrix elements need not be normally distributed. Mathematicians refer to this as *universality*. Similarly, physicists typically refer to phenomena as universal if they are widely observed regardless of the details or the particular model assumptions. As we have previously seen, the phenomenon of level repulsion captures a broad set of physical and mathematical realizations, which are surprisingly well captured by Wigner's surmise. This is an example of a *physical* universality – we often do not have solid mathematical proof of why the surmise works for these various cases. But despite the *mathematical* universality associated with the semicircle law, it is actually not exhibited for the eigenvalue distribution of matrices associated with most physical realizations – even in the case of the energy level of nuclei (which sparked the whole field of random matrix theory!), while the random matrix approach captures well the statistics of level spacings, the eigenvalue distribution has little to do with the semicircle law – since the Hamiltonian is of course not a set of i.i.d. random variables.

8.3 Joint Probability Distribution of Eigenvalues

Surprisingly, it is also possible to obtain the joint probability distribution of eigenvalues, albeit by a completely different method. We will see that this form may be used to recover the semicircle law, yielding a very different intuition for it, and will also manifest the level repulsion. The derivation generally follows that of Mehta (2004).

For the GOE, as we have seen earlier, the number of degrees of freedom (i.e., independent matrix entries) is

$$\gamma = N + (N^2 - N)/2 = N(N + 1)/2. \tag{8.33}$$

Consider now a mapping between these γ independent variables and a new set of γ variables, N of which will be the eigenvalues, and the additional $M = N(N - 1)/2$ variables – which we will explicitly specify later – will be denoted $P_1, P_2 .. P_M$. By diagonalizing the matrix H we find that

$$e^{-\frac{N}{4} tr H^2} = e^{-\frac{N}{4} \sum_i \lambda_i^2}. \tag{8.34}$$

Therefore, we have

$$p(\lambda_1, \lambda_2 \ldots \lambda_N) = \int e^{-\frac{N}{4} \sum_i \lambda_i^2} |det(J)| dP_1 \ldots dP_M, \tag{8.35}$$

where J is the Jacobian of the transformation.

The difficulty will be in determining the Jacobian. Consider now a particular choice of variables P_j, defined as follows: We will diagonalize the matrix H using a real, orthogonal matrix U, such that

$$H = UDU^t. \tag{8.36}$$

The diagonal matrix D has the eigenvalues of H on its diagonal, and we can choose them to be ordered by their value (denoting λ_j as the jth element on the diagonal). The rows of U are the corresponding eigenvectors, which are orthogonal to each other and have a norm of 1. The only freedom left in their definition is a choice of sign (we can multiply them by ± 1) – which can be removed by demanding that their first component would be positive. In this way we defined U and D uniquely, and clearly the mapping is invertible. You may indeed check that the number of degrees-of-freedom is the same, as must be the case for a one-to-one mapping: The matrix D has N of the γ degrees of freedom, and the matrix U has the rest $N(N - 1)/2$ (can you see why?), which we will name $P_1 .. P_M$. According to our mapping, taking the derivative of U with respect to any of the λ_j (while keeping the P's constant) is trivially 0. Consider the relation

$$U^t U = I. \tag{8.37}$$

Taking the derivative with respect to P_j gives us

$$\frac{\partial U^t}{\partial P_j} U + U^t \frac{\partial U}{\partial P_j} = 0. \tag{8.38}$$

Hence if we define

$$S^j \equiv \frac{\partial U^t}{\partial P_j} U, \tag{8.39}$$

Eq. (8.38) implies that S^j is an anti-symmetric matrix. Similarly,

$$\frac{\partial H}{\partial P_j} = \frac{\partial U}{\partial P_j} DU^t + UD\frac{\partial U^t}{\partial P_j}. \tag{8.40}$$

Multiplying by U^t to the left and U to the right we find that

$$U^t\frac{\partial H}{\partial P_j}U = U^t\frac{\partial U}{\partial P_j}D + D\frac{\partial U^t}{\partial P_j}U = DS^j - S^j D. \tag{8.41}$$

We can write this relation explicitly as

$$\sum_{l,m} U^t_{\alpha l}\frac{\partial H_{lm}}{\partial P_j}U_{m\beta} = [DS^j - S^j D]_{\alpha\beta}, \tag{8.42}$$

which can be rewritten as

$$\sum_{l,m} \frac{\partial H_{lm}}{\partial P_j}U_{l\alpha}U_{m\beta} = (\lambda_\alpha - \lambda_\beta)S^j_{\alpha\beta}. \tag{8.43}$$

Similarly, we can calculate the derivatives of H with respect to λ_j, using the fact that $\frac{\partial U}{\partial \lambda_j} = 0$, and find that

$$\frac{\partial H}{\partial \lambda_j} = U\frac{\partial D}{\partial \lambda_j}U^t. \tag{8.44}$$

Hence by multiplying with U^t on the left and U on the right we find that

$$U^t\frac{\partial H}{\partial \lambda_j}U = \frac{\partial D}{\partial \lambda_j}. \tag{8.45}$$

We have

$$\left[\frac{\partial D}{\partial \lambda_j}\right]_{\alpha\beta} = \delta_{\alpha j}\delta_{\beta j}, \tag{8.46}$$

therefore we can write Eq. (8.45) as

$$\sum_{l,m} \frac{\partial H_{lm}}{\partial \lambda_j}U_{l\alpha}U_{m\beta} = \delta_{\alpha j}\delta_{\beta j}. \tag{8.47}$$

Consider now the Jacobian that we are after, per Eq. (8.35):

$$J(\lambda's, P's) = \begin{pmatrix} \frac{\partial H_{11}}{\partial \lambda_1} & \cdots & \frac{\partial H_{\alpha\beta}}{\partial \lambda_1} & \cdots \\ \cdots & \cdots & \cdots & \cdots \\ \frac{\partial H_{11}}{\partial \lambda_N} & \cdots & \frac{\partial H_{\alpha\beta}}{\partial \lambda_N} & \cdots \\ \frac{\partial H_{11}}{\partial P_1} & \cdots & \frac{\partial H_{\alpha\beta}}{\partial P_1} & \cdots \\ \cdots & \cdots & \cdots & \cdots \\ \frac{\partial H_{11}}{\partial P_M} & \cdots & \frac{\partial H_{\alpha\beta}}{\partial P_M} & \cdots \end{pmatrix}, \tag{8.48}$$

i.e., the first N rows are the derivatives with respect to the λ's and the rest are with respect to the P's. For convenience, let us choose the first N columns to correspond to the diagonal elements of H.

Let us now define another $\gamma \times \gamma$ matrix, as follows: For $l = 1 \ldots N$, the lth row will be

$$V_l = [U_{l\alpha}U_{l\beta}], \tag{8.49}$$

where $\alpha \leq \beta$ (and both are integers between 1 and N – as will be the case throughout). We can characterize one of the rest $N(N-1)/2$ rows by two indices l, m with $l < m$, and it will be defined as

$$V_{l,m} = U_{l\alpha}U_{m\beta} + U_{m\alpha}U_{l\beta}. \tag{8.50}$$

We will choose the first N columns to correspond to $\alpha = \beta$, and the rest to $\alpha < \beta$.

Armed with this definition, let us compute the matrix JV. Taking the jth row of J, with $j \leq N$, and multiplying it with the column of V corresponding to the pair $\alpha \leq \beta$, we have

$$[JV]_{j,\{\alpha,\beta\}} = \sum_{l=1}^{N} \frac{\partial H_{ll}}{\partial \lambda_j} U_{l\alpha}U_{l\beta} + \sum_{l<m} \frac{\partial H_{lm}}{\partial \lambda_j}[U_{l\alpha}U_{m\beta} + U_{m\alpha}U_{l\beta}]. \tag{8.51}$$

Note that this sum is identical in structure to that of Eq. (8.47)! We therefore find that

$$[JV]_{j,\{\alpha,\beta\}} = \delta_{j\alpha}\delta_{j\beta}, \tag{8.52}$$

where the *column* of the matrix is characterized as before by a pair $\{\alpha, \beta\}$ with $\alpha \leq \beta$. This implies that the first N rows of JV consist of the identity matrix on the LHS, and zeros on the RHS.

Similarly, taking the $(j+N)$th row of J and multiplying it with the column of V corresponding to the pair $\alpha \leq \beta$, we have

$$[JV]_{j+N,\{\alpha,\beta\}} = \sum_{l=1}^{N} \frac{\partial H_{ll}}{\partial P_j} U_{l\alpha}U_{l\beta} + \sum_{l<m} \frac{\partial H_{lm}}{\partial P_j}[U_{l\alpha}U_{m\beta} + U_{m\alpha}U_{l\beta}]. \tag{8.53}$$

This sum is identical in structure to that of Eq. (8.43)! We therefore find that

$$[JV]_{j,\{\alpha,\beta\}} = (\lambda_\alpha - \lambda_\beta)S_{\alpha\beta}^j. \tag{8.54}$$

Therefore JV has the structure:

$$JV = \begin{pmatrix} 1 & 0 & 0 & & 0 & & 0 \\ 0 & .. & 0 & & 0 & & 0 \\ 0 & 0 & 1 & & 0 & & 0 \\ 0 & 0 & 0 & & & & \\ 0 & 0 & 0 & & S_{\alpha,\beta}^j(\lambda_\alpha - \lambda_\beta) & & \\ 0 & 0 & 0 & & & & \end{pmatrix}. \tag{8.55}$$

Now we are done with the technical part of the derivation, and close to our goal of finding the joint probability distribution. Note that the submatrix $S^j_{\alpha\beta}(\lambda_\alpha - \lambda_\beta)$ (with dimensions $\frac{N(N-1)}{2} \times \frac{N(N-1)}{2}$), can be written as the product of the two matrices:

$$A \equiv S^j_{\alpha\beta} \tag{8.56}$$

(with $\alpha < \beta$) and B a *diagonal* matrix whose diagonal elements are $(\lambda_\alpha - \lambda_\beta)$.

Taking the determinant of JV will therefore give us

$$det(JV) = \prod_{\alpha<\beta} |\lambda_\alpha - \lambda_\beta| det(A). \tag{8.57}$$

Importantly, from its definition the matrix S^j, and hence A, only depend on the P variables, and thus will not depend on the λ variables! Additionally, V only depends on the components of U, and hence $det(V)$ is independent of the λ variables, and we have

$$det(J) = \prod_{\alpha<\beta} |\lambda_\alpha - \lambda_\beta| det(A) f(P_1, P_2..P_M) = g(P_1, P_2..P_M) \prod_{\alpha<\beta} |\lambda_\alpha - \lambda_\beta|. \tag{8.58}$$

Since the integral over the Jacobian runs only over the P variables (see Eq. (8.35)), we conclude that

$$p(\lambda_1,..\lambda_N) \propto e^{-\frac{N}{4}\sum_i \lambda_i^2} \prod_{j<k} |\lambda_j - \lambda_k|. \tag{8.59}$$

Getting the constant of proportionality is possible (Livan *et al.* 2018) but we will not derive it here. The RHS side of this equation, coming from the Jacobian, is the level repulsion – we see that the probability distribution of finding nearby eigenvalues is vanishingly small.

Beyond the semicircle law: top eigenvalues In the last part of Chapter 6, the maximum of a large number of independent variables drawn from a particular distribution is considered (leading, surprisingly, to three universality classes). It is natural and interesting, therefore, to consider the case of *correlated* variables. One important scenario that has received much attention is the case of the top eigenvalue of a Hermitian random matrix, where eigenvalues are correlated as manifested explicitly in Eq. (8.59). In this case it can be shown that the typical deviation from the $\sqrt{2N}$ associated with the semicircle law scales as $N^{-1/6}$. When this scaling is accounted for, an interesting (N-independent) distribution known as the Tracy–Widom distribution emerges (Tracy and Widom 1994). Remarkably, this distribution has been shown to "pop up" in numerous and unexpected scenarios, including sequence alignments (Majumdar and Nechaev 2005) and models of surface growth (Kriecherbauer and Krug 2010), to name but. One may also consider the *tail* of the top eigenvalue distribution – going beyond the Tracy–Widom distribution – see Majumdar and Schehr (2014). Note that the same question for *non-Hermitian* matrices is discussed later in the box "Beyond the circular law."

8.3.1   From the Joint Probability Distribution to the Semicircle Law

Eq. (8.59) contains all the information regarding the eigenvalue statistics. We should therefore be able to get the semicircle law from it (as well as an exact form which replaces the approximation of Wigner's surmise). This can be done by evaluating a (difficult) high-dimensional integral, which is a path we will not pursue in this book. Instead, we will give an elegant derivation, once again due to Wigner (1957), that essentially maps this problem to one in statistical mechanics.

The density-of-states (DOS) measures the density of eigenvalues at a given point λ, and is given by

$$\sigma_N(\lambda) = N \int_{-\infty}^{\infty} \cdots \int_{-\infty}^{\infty} p(\lambda_1, \lambda_2 \ldots \lambda_N) \delta(\lambda - \lambda_1) d\lambda_1 \, d\lambda_2 \ldots d\lambda_N. \quad (8.60)$$

(Note that the DOS is proportional to the eigenvalue probability distribution we studied earlier, integrating to N rather than 1).

Consider a gas of particles which can move on a one-dimensional line, confined by a harmonic potential, and interacting with each other via a repulsive logarithmic interaction, $U(r) = -\log(|r|)$. Denoting the positions of the particles by λ_i, the energy of the system is given by

$$E = \sum_{i=1}^{N} \frac{1}{2} k \lambda_i^2 - \sum_{i<j} \log(|\lambda_i - \lambda_j|), \quad (8.61)$$

where k is the spring constant associated with the harmonic potential. If we choose $k \equiv N/2$, and put the particles in a thermal bath with $kT = 1$, then according to Boltzmann the probability to observe a configuration $\lambda_1 \ldots \lambda_N$ is given by

$$p(\lambda_1, \lambda_2 \ldots, \lambda_N) = e^{-E/kT} \propto e^{-\frac{N}{4} \sum \lambda_i^2 + \sum_{i<j} \log(|\lambda_i - \lambda_j|)}. \quad (8.62)$$

This is precisely the joint probability distribution of Eq. (8.59)! Soon we shall show that for large N, where we may work with a continuous particle (charge) distribution, the configuration which minimizes the energy functional of Eq. (8.61) is precisely the semicircle law. This would be the state of the system at *zero* temperature, but Eq. (8.60) shows that we need to find the average charge density at a finite temperature of $kT = 1$ in our statistical mechanics interpretation – why should the two be the same? The answer is that in the minimal energy configuration the energy per particle will scale as N, while the thermal energy will be of order unity – and hence will be negligible. Indeed, in the semicircle law the typical λ is of order unity, and hence $\frac{1}{2} N \lambda^2 = O(N)$ (and similarly, the interaction energy with the other N particles will be of order $O(N)$ as well). Therefore, we proceed to find the minimal energy configuration of the system, which will suffice to find the DOS. Note also that the physical system can be interpreted as N charged, parallel, infinite wires, forced to be in a given plane – the Coulomb interaction falling off as $1/r$ between charges leads to a logarithmic interaction between the charged wires. This is useful since it will allow us to use the intuition we developed for electrostatics, and in particular Gauss's law. This mapping is often referred to as Dyson's Coulomb gas.

In the large N limit, the minimal energy configuration can be described by a continuous charge density $\sigma(\lambda)$, and the energy is given by

$$E(\sigma) = \int \sigma(\lambda)\frac{1}{2}k\lambda^2 d\lambda - \frac{1}{2}\int\int \sigma(\lambda_i)\sigma(\lambda_j)\log(|\lambda_i - \lambda_j|)d\lambda_i \, d\lambda_j. \quad (8.63)$$

We must impose the constraint

$$\int_{-\infty}^{\infty} \sigma(\lambda) = N. \quad (8.64)$$

Minimizing the functional of Eq. (8.63) under this constraint can be done with Lagrange multipliers (see Appendix E for a reminder). Taking the functional derivative immediately gives us

$$\frac{1}{2}k\lambda^2 - \int \sigma(\lambda')\log(|\lambda - \lambda'|)d\lambda' = G, \quad (8.65)$$

where the constant G is the Lagrange multiplier.

It is tempting to take the derivative with respect to λ, but this is problematic since the second term would give

$$\int \sigma(\lambda')\frac{1}{\lambda - \lambda'}d\lambda', \quad (8.66)$$

which is not well defined. There is no problem with convergence of the integral of Eq. (8.65) though! For this reason we may replace the integration limits with

$$\lim_{\epsilon \to 0}\left[\int_{-\infty}^{\lambda-\epsilon} d\lambda' + \int_{\lambda-\epsilon}^{\infty} d\lambda'\right]. \quad (8.67)$$

Now we may take the derivative and find that

$$P\left(\int_{-\infty}^{\infty}\frac{\sigma(\lambda')}{\lambda - \lambda'}d\lambda'\right) = k\lambda, \quad (8.68)$$

where the "P" denotes taking the principal value of the integral, as defined in Eq. (8.67).

A subtle point is that this equation was derived by taking the functional derivative and equating it to zero, i.e., $\sigma \to \sigma + \delta\sigma$ should give vanishing contribution to first order. However, since σ is nonnegative we can only do this procedure for x where $\sigma(x) > 0$, i.e., if we find a solution with bounded support, the equation should not be fulfilled for x outside the support.

Eq. (8.68) is known as Tricomi's equation, and there are systematic methods of solving such equations (for an arbitrary function of λ on the RHS), see Tricomi (1985) and Muskhelishvili and Radok (2008). Alternatively, you may "guess" a solution (e.g., inspired by numerics), and prove that it solves the equation (though this relies on the assumption that the solution is unique). Plugging in the semicircle law we have

$$\sigma(x) = C\sqrt{4 - x^2}, \quad (8.69)$$

with $C = \frac{N}{2\pi}$. Performing the integral we find that

$$\int \frac{\sqrt{4 - y^2}}{x - y} dy = x \sin^{-1}(y/2) - \sqrt{4 - y^2} + \sqrt{4 - x^2}$$

$$\times \log\left(\sqrt{4 - x^2}\sqrt{4 - y^2} - xy + 4\right) - \sqrt{4 - x^2} \log|x - y|.$$

$$(8.70)$$

Plugging in the boundaries $(-2, x - \epsilon)$ and $(x + \epsilon, 2)$, we find that the contribution of the first term is $x\pi$, while that of the second term clearly vanishes. The last two terms can also be shown to give vanishing contribution when combined. Overall we find that

$$\frac{N}{2\pi}\pi x = xN/2 = kx, \qquad (8.71)$$

which is indeed a solution since $k = N/2$.

Dyson's Coulomb gas As noted above, the derivation above was originally performed by Eugene Wigner. However, several years later Freeman Dyson made an even more profound connection between RMT and the statistical mechanics of the Coulomb gas, in another seminal work (Dyson 1962). Dyson showed that as each of the matrix elements of the random matrix *independently* performs a random walk in a parabolic confining potential (think of the Langevin dynamics of Chapter 3, and in particular the Ornstein–Uhlenbeck process), the eigenvalues are also described by Langevin dynamics, albeit with an additional interaction term. The interaction falls off as the inverse of the separation of the eigenvalues, a result which comes naturally in this approach using (standard) perturbation theory.

8.4   Ensembles of Non-Hermitian Matrices and the Circular Law

While in the context of physics it is natural to consider *Hermitian* matrices, there is no reason to expect the relevant matrices to be Hermitian when dealing with, e.g., a matrix describing the dynamics of an ecological or neural network, as we shall soon see in a particular application dating back to the 1970s. To paraphrase a famous quote, calling a matrix non-Hermitian is akin to calling a fruit a non-banana ... What is the parallel to Wigner's semicircle law for ensemble of non-Hermitian matrices? Is there a parallel to level repulsion? Such problems are rather difficult to answer analytically, but surprisingly easy to test numerically. For example, consider the following *single* line of MATLAB code:

```
plot(eig(randn(900)),'.');
```

The code results in Fig. 8.5.

The eigenvalues appear to be uniformly distributed in a disk with radius $\sqrt{N} = 30$ (see also Problem 8.5, which explores this further numerically). This

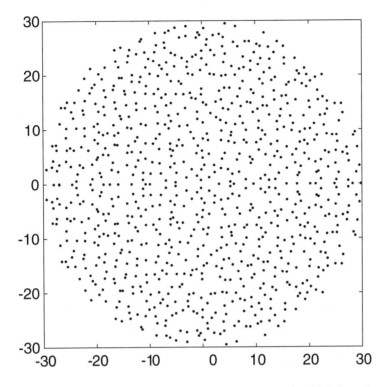

Figure 8.5 Eigenvalues of a 900×900 non-Hermitian matrix with independent, real entries drawn from a Gaussian distribution.

is precisely the content of the circular law, which we shall prove. First, let's define the matrix ensembles. As a reminder, for the GOE we had real, symmetric matrix with probability distribution

$$p(\boldsymbol{H}) \propto e^{-\frac{N}{4} tr \boldsymbol{H}^2}. \tag{8.72}$$

For the GUE, the ensemble consisted of complex, Hermitian matrices, with

$$p(\boldsymbol{H}) \propto e^{-\frac{N}{2} tr \boldsymbol{H}^2}. \tag{8.73}$$

Both result in the semicircle law. Consider now an ensemble of matrices with complex entries that are not necessarily Hermitian, with

$$p(\boldsymbol{H}) \propto e^{-tr \boldsymbol{H} \boldsymbol{H}^\dagger}. \tag{8.74}$$

This can be written as

$$p(\boldsymbol{H}) \propto e^{-\sum_{i,j} H_{ij} H_{ji}^\dagger} = e^{-\sum_{i,j} H_{ij} H_{ij}^*}. \tag{8.75}$$

Therefore, the real and imaginary parts of every matrix element are drawn independently from a Gaussian distribution.

Similarly, in an ensemble of real, non-Hermitian matrices, we can define

$$p(\boldsymbol{H}) \propto e^{-tr\boldsymbol{H}\boldsymbol{H}^t} = e^{-\sum_{i,j} H_{ij}^2}, \tag{8.76}$$

which is precisely what we simulated in the above code.

A beautiful result due to Jean Ginibre (1965) that we shall not prove is a formula for the *joint* probability distribution of the complex matrix ensemble of Eq. (8.75), stating that

$$p(\lambda_1, \lambda_2 \dots \lambda_N) \propto e^{-\sum_i |\lambda_i|^2} \prod_{i<j} |\lambda_i - \lambda_j|^2. \tag{8.77}$$

(In fact, the non-Hermitian ensemble we defined above is often referred to as the Ginibre ensemble).

Based on what we learnt, this result is plausible: The first term arises from the definition of the Gaussian ensemble, and the second will come from the relevant Jacobian, and corresponds to the level repulsion. The fact that the ensemble is of complex matrices implies that the level repulsion scales as s^2, rather than linearly.

Joint probability distribution for the real Ginibre ensemble Perhaps surprisingly, the result for the real Ginibre ensemble is not simply that of Eq. (8.77) with the level repulsion scaling linearly instead of quadratically (as is the case for the GOE vs. the GUE). The joint probability distribution for the real Ginibre ensemble is more involved, as was noticed already in Ginibre's (1965) seminal paper. In fact, we can see hints that things should be different already from the numerical simulation of Fig. 8.5, where a careful reader can notice a small fraction of eigenvalues precisely on the real axis. Eq. (8.77) would predict a negligible probability for eigenvalues to be *precisely* on the real axis. Note, however, that despite these subtleties for large N the fraction of the real eigenvalues vanishes and the circular law is recovered. For the exact distribution of eigenvalues of the real Ginibre ensemble, see Lehmann and Sommers (1991).

Using this result, we can easily derive the circular law, based on an analogy with electrostatics: Following the same logic that we used for the Hermitian case, we can map the problem to finding the lowest energy configuration of a system of interacting particles. As before, the $\sqrt{N}$ scaling of the semicircle law (that we have previously compensated for by the $\frac{N}{2}$ scaling in the definition of the matrix ensemble) implies that the large N case is, effectively, a low temperature limit – hence finding the lowest energy configuration suffices. In this case the "charges" live in the complex plane, and we may think of each of them as an infinite wire with a constant charge density, oriented *perpendicularly* to the complex plane. The repulsive interaction is logarithmic, as before, and the factor of 2 in Eq. (8.77) means that it is stronger by a factor of 2. The first, Gaussian, term is a harmonic trap which confines the particles to the vicinity of the origin (Note that this time the harmonic potential is proportional to $x^2 + y^2$).

So the problem boils down to the following: Which configuration of charges will yield an electric field which counteracts that of the harmonic trap, i.e., it should point out radially at every point and have magnitude linear in r? The numerics of Fig. 8.5 motivates us to guess a solution where the wires are uniformly distributed in a disk. Indeed, this implies that in the 3D space we have a uniform charge distribution in an infinite *cylinder*. From symmetry the electric field has to point in the radial direction, and applying Gauss's law (see Appendix E) to a cylindrical surface at distance r yields the equation

$$2\pi r L E(r) = \pi r^2 L \rho \implies E(r) \propto r, \qquad (8.78)$$

with ρ the charge density and $E(r)$ the electric field. Note that this is only correct for $r < R$, where R is the radius of the disk, but as in our derivation of the semicircle law the equation only has to be obeyed on the support of the DOS, and hence the different behavior in the region outside the disk is not an issue.

Universality of the circular law In fact, the circular law holds more generally than for the Gaussian ensembles discussed above (again, the mathematical universality we discussed in the context of Wigner's semicircle law also holds here). In the more general case, the radius of the disk scales as $I\sqrt{N}$, where I is the standard deviation of the i.i.d. matrix elements.

8.4.1 Application of the Circular Law to the Stability of Ecological Networks

Following May (1972), we will now use this result in a simple model for ecological networks. Consider a stable ecosystem comprising N species. Denoting the number of species of the jth type by n_j, we may attempt to write an equation for the dynamics of the system as

$$\frac{dn_i}{dt} = f(n_1, n_2..n_N), \qquad (8.79)$$

where $f(\vec{n})$ is a potentially complicated function describing the interactions of the species with each other. Let us assume that there exists a stable configuration in which the numbers fluctuate around the values $n_1^0, n_2^0..n_N^0$ (i.e., no limit cycles or other complex behavior). In this case we can *linearize* the equations of motion close enough to the fixed point, and obtain an equation of the form

$$\frac{d\vec{\delta n}}{dt} = A\vec{\delta n}, \qquad (8.80)$$

with A a (constant) matrix and $\vec{\delta n}$ the vector of deviations from the fixed point. Note that there is no constant term in the expansion since we are working near a fixed point.

Linear stability analysis We will next invoke the important concept of *linear stability analysis*. Taking a nonlinear system, we linearize the equations sufficiently close to a stable configuration (possibly one of many). We next ask the question: Suppose we perturb the system by an infinitesimal amount in some arbitrary direction. Will the perturbation dampen over time, or, alternatively, explode out of control? To answer this, we resort to the way of thinking utilized in Chapter 2, in the context of the Markov chain analysis: Since we have a set of linear equations, we can typically decompose the perturbation vector in terms of the eigenvectors of the matrix A of Eq. (8.80). Now you may readily see that the amplitude of each eigenvector grows *exponentially* with time, with a rate equal to the eigenvalue. Hence, the imaginary parts of the eigenvalues are associated with oscillations, while the real parts have to be negative for *all* eigenvalues in order for the system to be stable: otherwise, the amplitude of the corresponding eigenvector will *increase* exponentially over time (until the linear approximation fails, that is). For some examples, see Strogatz (2018).

May's idea – in a sense similar to Wigner's approach – was to replace A, for a complex, ecological system, by a random matrix with independent matrix entries, with typical magnitude I, accounting for the positive or negative effects that a given species may have on another (though unlike Wigner's Hermitian matrices describing quantum-mechanical systems, in this context there is no reason for the matrix to be Hermitian!). In addition, he postulated that the diagonal of the matrix consists of constant, negative elements, $-a$, which stabilize the system (i.e., a given species left on its own will be stable). The question is now: Is the system stable, in the sense of a linear stability analysis? For this to be true, the real part of all the eigenvalues has to be negative. Without the diagonal terms, the matrix A will have a form similar to the one we simulated above, and hence the eigenvalues will form a uniform, circular, cloud with radius $I\sqrt{N}$. This is clearly unstable – roughly half the eigenvalues will be negative. Due to the assumption of a *constant* element on the diagonal, the eigenvectors of the full matrix (with the diagonal) are unchanged, but the eigenvalues are shifted to the left by an amount a. For the system to be stable, we need a to be large enough to push all eigenvalues to the left of the y axis, hence we find that

$$I\sqrt{N} < a. \tag{8.81}$$

This implies that for a fixed a and I, there is a bound to the potential complexity of the ecological network! While there are many assumptions in this derivation that may be questioned, this approach has been highly influential and many works followed up on May's idea.

8.4.2   Interpolating the Semicircle Law and the Circular Law

[Note: the rest of this chapter may be skipped without affecting the flow of the rest of the text]

Sommers (1988) defines a matrix ensemble with a tunable parameter τ that interpolates the Hermitian ensembles with the non-Hermitian ones (known also as the "elliptical ensemble," for reasons we will shortly see). They define:

$$p(\boldsymbol{H}) \propto e^{-\frac{N}{2(1-\tau^2)} tr(\boldsymbol{H}\boldsymbol{H}^t - \tau \boldsymbol{H}^2)}. \tag{8.82}$$

Clearly, if $\tau = 0$ we are back to the non-Hermitian ensemble of Eq. (8.76). But it is less obvious why $\tau \to 1$ retrieves the Hermitian case. To see this, let us consider the correlations between pairs of matrix elements. Since we have

$$tr(\boldsymbol{H}\boldsymbol{H}^t - \tau \boldsymbol{H}^2) = \sum_{i,j} H_{ij}^2 - \tau H_{ij} H_{ji}, \tag{8.83}$$

clearly only pairs H_{ij} and H_{ji} are correlated (for $\tau \neq 0$). For a Gaussian probability distribution

$$p(\vec{v}) \propto e^{-\sum_{i,j} \frac{W_{ij}}{2} v_i v_j}, \tag{8.84}$$

we have

$$\langle v_i v_j \rangle = W_{ij}^{-1}. \tag{8.85}$$

To see this, go to a basis in which $\boldsymbol{W}$ is diagonal (we may choose $\boldsymbol{W}$ to be a symmetric matrix without loss of generality). We have

$$\boldsymbol{W} = \boldsymbol{U} \boldsymbol{D} \boldsymbol{U}^{-1}; \vec{\tilde{v}} = \boldsymbol{U} \vec{v}, \tag{8.86}$$

with $\boldsymbol{U}$ a unitary matrix. In fact, since $\boldsymbol{W}$ is real the matrix $\boldsymbol{U}$ can be chosen to be orthogonal (see Appendix B for a summary of these results from linear algebra). We therefore have (can you see why?)

$$p(\vec{\tilde{v}}) \propto e^{-\sum_i \frac{D_{ii}}{2} \tilde{v}_i^2}. \tag{8.87}$$

In the new basis, the different elements are uncorrelated, and we have

$$\langle \tilde{v}_i \tilde{v}_j \rangle = \delta_{ij} / D_{ii} = [\boldsymbol{D}^{-1}]_{ij}, \tag{8.88}$$

with $\boldsymbol{D}^{-1}$ the inverse of the (diagonal) matrix $\boldsymbol{D}$. Going back to the original basis leads to Eq. (8.85).

For the case of Eq. (8.83), for $i \neq j$ we have to compute the inverse of the following 2×2 matrix:

$$A = \begin{pmatrix} C & -C\tau \\ -C\tau & C \end{pmatrix}, \tag{8.89}$$

with $C = \frac{N}{1-\tau^2}$. Using Eq. (8.85) we find that

$$\langle H_{ij} H_{ij} \rangle = 1/N, \langle H_{ij} H_{ji} \rangle = \tau/N. \tag{8.90}$$

Therefore, as $\tau \to 1$ the paired off-diagonal matrix elements become perfectly correlated and the matrix approaches a Hermitian one. Note that for $\tau \neq 0$ the statistics are slightly different for the diagonal elements, but this has negligible effect on the eigenvalue distribution as $N \to \infty$. Similarly, Problem 8.6 shows that for $\tau \to -1$ the matrix becomes anti-symmetric, and the semicircle law is recovered – but this time when the eigenvalues are projected onto the imaginary axis.

8.4.3   Deriving the Circular Law

An interesting result of Sommers (1988) is that within the generalized ensemble that we defined in Eq. (8.82), the eigenvalues will be uniformly distributed in an *ellipse*. In the non-Hermitian limit this will recover the circular law, while for $\tau \to 1$ the ellipse will become flat and we will recover the semicircle law (since we will be effectively projecting the thin ellipse onto the real axis). We can test this result numerically, using the following code:

```
S=0.2;
N=1000;
B=randn(N);
A=(B+S*transpose(B))/2;
plot(eig(A),'.'); axis equal;
```

Note that the parameter S is related to τ in the previous formulation: for $S = 0$ the matrix is non-Hermitian with i.i.d. entries, while for $S = 1$ the matrix is Hermitian. Computing the Pearson correlation coefficient of the off-diagonal elements, it is easy to see that $\frac{2s}{1+s^2} = \tau$. Running the code results in Fig. 8.6, suggesting that the support of the eigenvalues is indeed an ellipse.

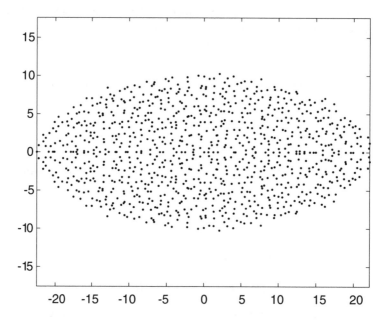

Figure 8.6 Eigenvalues of a 1000×1000 non-Hermitian matrix corresponding to Eq. (8.82).

To prove this result, which generalizes the circular law, it is useful to define the Green function $G(\omega)$,

$$G(\omega) = \frac{1}{N}\left[tr\left(\frac{1}{I\omega - H}\right)\right]_H, \tag{8.91}$$

where the $[\,]_H$ notation denotes ensemble-averaging over the matrix ensemble, using the definition of Eq. (8.82). In terms of the eigenvalues of H this can be written as

$$G(\omega) = \frac{1}{N}\left[\sum_{i=1}^{N}\frac{1}{\omega - \lambda_i}\right]_H. \tag{8.92}$$

Expressing this using the distribution of eigenvalues in the complex plane, $\rho(\lambda)$ (also known as the density-of-states, or DOS) we find that

$$G(\omega) = \int \frac{\rho}{\omega - \lambda}d^2\lambda, \tag{8.93}$$

where the integration is over the complex plane (i.e., over the real and imaginary parts of λ).

Our goal in this calculation is to find $\rho(\lambda)$. To see why that information is encoded in $G(\omega)$, consider a contour integral in the complex plane along a contour C, which encloses some domain R in the complex plane. Since $G(\omega)$ is a meromorphic function containing only simple poles, we have

$$I = \frac{1}{2\pi i}\oint G(\omega)d\omega = \sum_{\lambda \in R}1 = \int d^2\lambda \rho, \tag{8.94}$$

i.e., every eigenvalue in the domain R will contribute 1 to the sum. Defining $\omega = x + iy$, since $dy - idx = (dx + idy)/i$ we have

$$I = \frac{1}{2\pi}\oint_C (Gdy - iGdx). \tag{8.95}$$

Defining a vector field $\vec{D} \equiv (G, iG)$, this can also be interpreted as the flux of the vector field $\vec{D}$ through the boundary of the contour C: the normal to the surface points in the direction $d\vec{S} = (dy, -dx)$, and hence

$$\frac{1}{2\pi}\int \vec{D} \cdot d\vec{S} = I. \tag{8.96}$$

According to Gauss's theorem (see Appendix E), we have

$$I = \frac{1}{2\pi}\int_R \nabla \cdot \vec{D}d^2\lambda = \frac{1}{2\pi}\int_R \left(\frac{\partial G}{\partial x} + i\frac{\partial G}{\partial y}\right)d^2\lambda = \int_R d^2\lambda\rho. \tag{8.97}$$

Taking the imaginary part, this proves that

$$Im\left[\frac{\partial G}{\partial x} + i\frac{\partial G}{\partial y}\right] = 0, \tag{8.98}$$

which will enable us to derive an electrostatic analogy, where given G we can define a potential and electric field. From Eq. (8.98) we have

$$\frac{\partial ImG}{\partial x} = -\frac{\partial ReG}{\partial y}. \tag{8.99}$$

Defining

$$E_x \equiv 2ReG; E_y \equiv -2ImG, \tag{8.100}$$

we have

$$\frac{\partial E_x}{\partial y} - \frac{\partial E_y}{\partial x} = 0. \tag{8.101}$$

Hence, $\vec{E}$ can be derived from a potential:

$$\vec{E} = -\nabla\phi(x, y). \tag{8.102}$$

We can also write the DOS in terms of the Laplacian of the potential:

$$-\nabla^2\phi = \nabla \cdot \vec{E} = 2\frac{\partial ReG}{\partial x} - 2\frac{\partial ImG}{\partial y} = 2Re\left[\frac{\partial G}{\partial x} + i\frac{\partial G}{\partial y}\right] = 4\pi\rho. \tag{8.103}$$

Hence the potential can be interpreted as the electrostatic potential arising from a charge distribution with charge density corresponding to the DOS. We will now show that defining

$$\phi(\omega) = -\frac{1}{N}\left[\log[det(I\omega^* - H^t)(I\omega - H)]\right]_H \tag{8.104}$$

is consistent with Eq. (8.103). To prove this, consider a basis where H is diagonal. We then have

$$det(I\omega - H) = \prod_{j=1}^{N}(\omega - \lambda_j); det(I\omega^* - H^t) = \prod_{j=1}^{N}(\omega^* - \lambda_j), \tag{8.105}$$

hence

$$\phi = -\frac{1}{N}\left[\log\left[\prod_{j=1}^{N}(\omega - \lambda_j)(\omega^* - \lambda_j)\right]\right]_H. \tag{8.106}$$

$$\implies -\frac{\partial\phi}{\partial x} = \frac{1}{N}\left[\sum_j\left[\frac{1}{\omega - \lambda_j} + \frac{1}{\omega^* - \lambda_j}\right]\right]_H. \tag{8.107}$$

and

$$-\frac{\partial \phi}{\partial y} = \frac{1}{N} \left[\sum_j \left[\frac{i}{\omega - \lambda_j} - \frac{i}{\omega^* - \lambda_j} \right] \right]_H . \tag{8.108}$$

$$\implies -\left(\frac{\partial \phi}{\partial x} - i \frac{\partial \phi}{\partial y} \right) = \frac{2}{N} \left[\sum \frac{1}{\omega - \lambda_j} \right]_H = 2G(\omega), \tag{8.109}$$

as is indeed implied from Eqs. (8.100) and (8.103).

A central part to the derivation is that we shall now replace the order of the ensemble-averaging and the logarithm, i.e., we will compute

$$\tilde{\phi}(\omega) = \frac{1}{N} \log \left(\left[det^{-1} (I\omega^* - H^t)(I\omega - H) \right]_H \right) . \tag{8.110}$$

It can be proven that this non-trivial replacement does not change the result in the large N limit, by using the "replica-trick," first introduced in the context of spin-glasses. The idea is to use the formula:

$$\log Z = \lim_{n \to 0} \frac{Z^n - 1}{n}, \tag{8.111}$$

compute $\langle Z^n \rangle$ for *integer* values of n (when the calculation is doable), and then analytically continue the dependence of the RHS of Eq. (8.111) to find the $n \to 0$ limit and obtain $\langle \log Z \rangle$. We will not go through this procedure, and simply use the result from it, namely, that we can ensemble-average det^{-1} and then take the logarithm rather than compute the ensemble-average of $-\log[det]$. Note that this is not always the case – in fact, it is quite surprising that making this replacement does not change the result! This occurs since the distribution of the matrix determinant essentially becomes a δ-function, for which the expectation value of the logarithm of the variable is identical to the logarithm of the expectation value (it is quite easy to corroborate this fact numerically).

To proceed, we need to calculate

$$Q \equiv \left[det^{-1} (I\omega^* - H^t)(I\omega - H) \right]_H . \tag{8.112}$$

In general, for a Hermitian, positive-definite, matrix A_{ij}, we may replace $1/det(A)$ by the following integral:

$$1/det(A) = \int \prod_m \frac{d^2 z_m}{\pi} e^{-\sum_{i,j} A_{ij} z_i^* z_j} \tag{8.113}$$

(where the integral is over the real and imaginary parts of each of the z_m complex variables.) To see this, we can diagonalize A using a unitary matrix and find that

$$A = U^\dagger D U; z_i \to \tilde{z}_i, \tag{8.114}$$

leading to

$$\sum_{i,j} A_{ij} z_i^* z_j = \sum_{ijkl} U_{ki}^* D_{kl} U_{lj} z_i^* z_j = \sum_{ijk} \lambda_k U_{ki}^* z_i^* U_{kj} z_j = \sum_k \lambda_k \tilde{z}_k^* \tilde{z}_k. \quad (8.115)$$

Since U is unitary the Jacobian is 1, and we have

$$\int \prod_m \frac{d^2 z_m}{\pi} e^{-\sum_{i,j} A_{ij} z_i^* z_j} = \int \prod_m \frac{d^2 \tilde{z}_m}{\pi} e^{-\sum_k \lambda_k (Re[\tilde{z}_k]^2 + Im[\tilde{z}_k]^2)} = \frac{1}{\prod_k \lambda_k} = 1/det(A).$$

$$(8.116)$$

In our case the structure of Eq. (8.112) ensures that all eigenvalues are nonnegative, but they can be zero. Nonetheless, we can replace $1/det(A)$ with the following limit:

$$Q = \lim_{\epsilon \to 0} \left[\int \prod_m \frac{d^2 z_m}{\pi} e^{-\epsilon \sum_i |z_i|^2} e^{-\sum_{ijk} z_i^* (\omega^* \delta_{ik} - H_{ik}^\dagger)(\omega \delta_{kj} - H_{kj}) z_j} \right]_H. \quad (8.117)$$

As we will see later, this subtlety will have important consequences! (ϵ will be a regularizer, reminiscent of the $\pm i\epsilon$ regularization often used in the context of Green's functions calculations.) Now we can perform the ensemble average using Eq. (8.82). For simplicity, we will proceed under the assumption $\tau = 0$ (corresponding to the circular law). Performing the ensemble-averaging in Eq. (8.117) implies integration over N^2 variables, but in this case since H_{ki} only couples to terms of the form H_{kj}, the integral decouples into a product of N integrals. To evaluate each of them we can fix k, and define $v_j \equiv H_{kj}$. Consider the integration over the N variables $v_1, v_2 \ldots v_N$. Writing the integral explicitly – aside from the ϵ related term which we can factor out of the integral – we have to compute

$$I_k = \int dv_1 dv_2 \ldots dv_N e^{-\sum_{i,j} A_{ij} v_i v_j - \sum_j b_j v_j} e^{-|\omega|^2 |z_k|^2} \quad (8.118)$$

with the quadratic form

$$A_{ij} \equiv z_i^* z_j + \frac{N}{2} \delta_{ij} \quad (8.119)$$

(where the second term arises from the definition of the matrix ensemble) and the vector $\vec{b}$ is

$$b_j \equiv \omega^* z_k^* z_j + C.C. \quad (8.120)$$

The factor $e^{-|\omega|^2 |z_k|^2}$ can be taken out of the integral. To evaluate the integral over the remaining terms, we can proceed by completing the square, as we have done several times before in computing Gaussian integrals. First, it would be more convenient to work with real rather than complex quantities, and we may readily rewrite the quadratic form as

$$\sum_{i,j} A_{ij} v_i v_j = \sum_{i,j} \left(z_i' z_j' + z_i'' z_j'' + \frac{N}{2} \delta_{ij} \right) v_i v_j, \quad (8.121)$$

where in this derivation we will (occasionally) be using x' and x'' to denote the real and imaginary parts of a variable x, respectively. We will denote the symmetrized quadratic form as

$$B_{ij} \equiv z_i' z_j' + z_i'' z_j'' + \frac{N}{2} \delta_{ij}. \tag{8.122}$$

It will be helpful to rewrite the expression in Eq. (8.118) as

$$I_k = P \int dv_1 dv_2 \ldots dv_N e^{-\sum_{i,j} B_{ij}(v_i - c_i)(v_j - c_j)} e^{-|\omega|^2 |z_k|^2}, \tag{8.123}$$

with $P = e^{\sum_{i,j} B_{ij} c_i c_j}$ (as usual when completing the square in Gaussian integrals). Clearly, the terms quadratic in v_j are identical to those of Eq. (8.118). If we demand that the linear components also match, we obtain the following equation:

$$\sum_j 2 B_{ij} c_j = b_j. \tag{8.124}$$

Fortunately, it is not hard to solve this. Guessing a solution of the form $c_j = a_1 z_j' + a_2 z_j''$, we find that

$$\sum_j B_{ij} c_j = \frac{N}{2} [a_1 z_i' + a_2 z_i''] + \sum_i [z_i' z_j' + z_i'' z_j''][a_1 z_j' + a_2 z_j'']. \tag{8.125}$$

Throughout this derivation we will use the following three important variables:

$$S_1 \equiv \sum_i (Re[z_i])^2; \; S_2 \equiv \sum_i (Im[z_i])^2; \; S_3 \equiv \sum_i Re[z_i] Im[z_i]. \tag{8.126}$$

Using this notation, we have

$$\sum_i [z_i' z_j' + z_i'' z_j''][a_1 z_j' + a_2 z_j''] = a_1 [S_1 z_i' + S_3 z_i''] + a_2 [S_3 z_i' + S_2 z_i''], \tag{8.127}$$

while

$$b_i / 2 = Re[\omega^* z_k^* z_i] = \omega'(z_k' z_i' + z_k'' z_i'') + \omega''(z_k' z_i'' - z_k'' z_i'). \tag{8.128}$$

Equating the coefficients of z_j' and z_j'' in Eqs. (8.125) and (8.128) we find that

$$\left[\frac{N}{2} + S_1 \right] a_1 + S_3 a_2 = \omega' z_k' - \omega'' z_k'' \equiv C_1, \tag{8.129}$$

$$S_3 a_1 + \left[\frac{N}{2} + S_2 \right] a_2 = \omega' z_k'' + \omega'' z_k' \equiv C_2. \tag{8.130}$$

Solving, we find that

$$a_1 = \frac{1}{[\frac{N}{2} + S_1][\frac{N}{2} + S_2] - S_3^2} \left(\left[\frac{N}{2} + S_2 \right] C_1 - S_3 C_2 \right), \tag{8.131}$$

$$a_2 = \frac{1}{[\frac{N}{2} + S_1][\frac{N}{2} + S_2] - S_3^2} \left(\left[\frac{N}{2} + S_1 \right] C_2 - S_3 C_1 \right). \tag{8.132}$$

Going back to the integral of Eq. (8.123), clearly the shifts by c_i do not affect the result of the integration, and the Gaussian integral can be readily evaluated. Putting everything together, we find that the integral of Eq. (8.118) gives

$$I_k = \frac{\pi^{N/2}}{\sqrt{det(B)}} e^{\sum_{i,j} B_{ij} c_i c_j} e^{-|\omega|^2 |z_k|^2}. \tag{8.133}$$

We will now proceed to evaluate the determinant, to which end it will be instructive to determine its eigenvectors and eigenvalues. Considering the form of Eq. (8.122), the $N/2$ terms on the diagonal clearly only shift all eigenvalues by $N/2$. We shall denote the matrix *without* the $N/2$ terms by $\tilde{B}$. We may use the following ansatz for its eigenvectors:

$$t_i = Re[z_i] + x Im[z_i], \tag{8.134}$$

where x is to be determined. A straightforward calculation shows that this will be an eigenvector provided that x obeys the quadratic equation

$$x^2 S_3 + x(S_1 - S_2) - S_3 = 0. \tag{8.135}$$

The corresponding eigenvalues are $\lambda = S_1 + x S_3$, given explicitly by

$$\lambda_\pm = \frac{S_1 + S_2}{2} \pm \frac{1}{2} \sqrt{(S_1 - S_2)^2 + 4 S_3^2}. \tag{8.136}$$

The sum of these two eigenvalues happens to be equal to the trace of the matrix $\tilde{B}$. It is easy to see that the quadratic form $\sum_{i,j} \tilde{B}_{ij} x_i x_j$ cannot be negative (can you see why?) implying that all other eigenvalues cannot be negative – hence they must vanish! Thus the determinant of the matrix B equals

$$det(B) = \left(\frac{N}{2}\right)^{N-2} \left(\lambda_+ + \frac{N}{2}\right) \left(\lambda_- + \frac{N}{2}\right). \tag{8.137}$$

Note that since we only care about the potential ϕ up to an additive constant, the proportionality constant will not matter – we only need to account for terms which depend on the z_j variables or ω.

Next, we will evaluate the contribution of the terms $e^{-\sum_{i,j} B_{ij} c_i c_j} e^{-|\omega|^2 |z_k|^2}$ of Eq. (8.133). Since we are interested in the product $I \equiv \prod_k I_k$, we need to evaluate

$$I = \left[\frac{\pi^N}{det(B)}\right]^{\frac{N}{2}} e^{-|\omega|^2 \sum_k |z_k|^2} e^{-\sum_{i,j,k} B_{ij} c_i c_j}. \tag{8.138}$$

(Note that while the matrix B does not depend explicitly on k, the elements c_i do.)

Within our notation $\sum_k |z_k|^2 = S_1 + S_2$. Evaluating the second sum is straightforward but more tedious. Considering B in Eq. (8.122), the contribution of the $\frac{N}{2} \delta_{ij}$ is

$$X = \sum_k \frac{N}{2} \sum_i c_i^2 = \sum_k \sum_i (a_1 z_i' + a_2 z_i'')^2 = \frac{N}{2} \sum_k [a_1^2 S_1 + a_2^2 S_2 + 2 a_1 a_2 S_3]. \tag{8.139}$$

The only dependence on k enters through a_1 and a_2. Plugging in the expression for a_1 and performing the summation over k we find that

$$\sum_k a_1^2 = \frac{1}{T^2} \left(\left[\frac{N}{2} + S_2 \right]^2 \sum_k C_1^2 + S_3^2 \sum_k C_2^2 - 2S_3 \left[\frac{N}{2} + S_2 \right] \sum_k C_1 C_2, \right)$$
(8.140)

with $T \equiv [\frac{N}{2} + S_1][\frac{N}{2} + S_2] - S_3^2$, and analogous expressions can be derived for $\sum_k a_2^2$ and $\sum_k a_1 a_2$:

$$\sum_k a_2^2 = \frac{1}{T^2} \left(\left[\frac{N}{2} + S_1 \right]^2 \sum_k C_2^2 + S_3^2 \sum_k C_1^2 - 2S_3 \left[\frac{N}{2} + S_1 \right] \sum_k C_1 C_2, \right),$$
(8.141)

$$\sum_k a_1 a_2 = \frac{1}{T^2} \left(-S_3 \left[\frac{N}{2} + S_2 \right] \sum_k C_1^2 - S_3 \left[\frac{N}{2} + S_1 \right] \right.$$
$$\left. \times \sum_k C_2^2 + \left(S_3^2 + \left[\frac{N}{2} + S_1 \right] \left[\frac{N}{2} + S_2 \right] \right) \sum_k C_1 C_2 \right). \quad (8.142)$$

Furthermore, we have

$$\sum_k C_1^2 = \omega'^2 S_1 + \omega''^2 S_2 - 2\omega'\omega'' S_3,$$
(8.143)

$$\sum_k C_2^2 = \omega'^2 S_2 + \omega''^2 S_1 + 2\omega'\omega'' S_3,$$
(8.144)

$$\sum_k C_1 C_2 = (\omega'^2 - \omega''^2) S_3 + \omega'\omega''(S_1 - S_2).$$
(8.145)

Finally, we need to account for the first two terms of B as defined in Eq. (8.122). This leads to a sum of the structure,

$$\sum_{i,j,k} [z_i' z_j' + z_i'' z_j''] c_i c_j = \sum_{i,j,k} [z_i' z_j' + z_i'' z_j''] (a_1 z_i' + a_2 z_i'')(a_1 z_j' + a_2 z_j''), \quad (8.146)$$

where as before the k dependence enters via the coefficients a_1, a_2. In an analogous fashion we may first fix k and perform the sum over i and j, resulting in the expression

$$S_k = a_1^2 (S_1^2 + S_3^2) + a_2^2 (S_2^2 + S_3^2) + 2a_1 a_2 (S_1 S_3 + S_2 S_3).$$
(8.147)

Performing the sum over k follows the same procedure as used previously. Putting it all together we have succeeded in precisely computing the integral and expressing it in terms of the real and imaginary parts of ω, as well as the three variables S_1, S_2, and S_3, which are defined in terms of the variables z_i!

Combining this with our formula for the determinant and the additional term $e^{-|\omega|^2(S_1+S_2)}$, we obtain after some algebra:

$$I = \left[\frac{\pi^N}{det(\boldsymbol{B})}\right]^{\frac{N}{2}} e^{-\frac{N\left(4(S_1 S_2 - S_3^2)\omega'^2 + 2(S_1^2 + S_2^2 + 2S_3^2)\omega''^2 + N|\omega|^2(S_1+S_2)\right)}{(2S_1+N)(2S_2+N)-4S_3^2}}. \tag{8.148}$$

A good sanity check of this expression is the case $N = 1$: then from Eq. (8.126) it is clear that $S_3 = \sqrt{S_1 S_2}$, upon which the expression for $det(\boldsymbol{B})$ simplifies to $[S_1 + S_2 + 1/2]$, and thus that of I to

$$I_{N=1} = \frac{\sqrt{2\pi}}{\sqrt{2(S_1 + S_2) + 1}} e^{-\frac{\left(2(S_1+S_2)^2\omega''^2 + |\omega|^2(S_1+S_2)\right)}{2(S_1+S_2)+1}}. \tag{8.149}$$

Performing the one-dimensional integral of Eq. (8.117) directly it is easy to verify this result.

Finally, to evaluate Q we have to go back to Eq. (8.117) and integrate over the z variables. To gain intuition regarding which region of the $2N$ dimensional space spanned by z_i' and z_i'' dominates, consider the N-dimensional vector $\vec{z}'$ whose elements are z_i' and analogously $\vec{z}''$ whose elements are z_i''. $S_{1,2}$ are respectively the squared norm of these two vectors, while $S_3 = \sqrt{S_1 S_2} \cos(\phi)$, where ϕ is the angle between these two vectors. Let us fix z_i' (and thus S_1) and consider the integration over the z_i''. We may perform the N-dimensional integral in spherical coordinates, choosing one of the axis to coincide with the direction of the vector $\vec{z}'$. The integral is proportional to

$$\int \tilde{r}^{N-1} \sin^{N-2}(\phi) I(S_1, S_2, S_3) d\tilde{r} d\phi, \tag{8.150}$$

with $S_2 = \tilde{r}^2$ and $S_3 = \sqrt{S_1 S_2} \cos(\phi)$. The term $\sin^{N-2}(\phi)$ strongly penalizes the vectors $\vec{z}'$ and $\vec{z}''$ from being parallel to each other: In the high-dimensional space this would be a limiting condition and it is favorable for the vectors to be orthogonal to each other. Writing the integrand as $e^{-Nf(\phi)}$, one can easily check that for sufficiently large N the function f is minimized when $\phi = \pi/2$, at which point $S_3 = 0$. Since we will be interested in sending N to infinity and evaluating the integral using a saddle-point approximation, we may thus go back to Eq. (8.148) and set $S_3 = 0$. Then I reduces to

$$I = \left[\frac{\pi^N}{det(\boldsymbol{B})}\right]^{\frac{N}{2}} e^{-\frac{N\left(4(S_1 S_2)\omega'^2 + 2(S_1^2 + S_2^2)\omega''^2 + N|\omega|^2(S_1+S_2)\right)}{(2S_1+N)(2S_2+N)}}. \tag{8.151}$$

Since the expression is symmetric with respect to S_1 and S_2, it is plausible that its maximum will occur when $S_1 = S_2$. In the case $S_3 = 0$, $S_1 = S_2 \equiv \frac{Nr}{2}$, where we have defined $r = \frac{1}{N}\sum_j |z_j|^2$. Then Eq. (8.151) reduces to

$$I = \left[\frac{\pi^N}{det(\boldsymbol{B})}\right]^{\frac{N}{2}} e^{-N|\omega|^2 \frac{r}{1+r}}. \tag{8.152}$$

Furthermore, in this case the determinant of $\boldsymbol{B}$ simplifies to

$$det(\boldsymbol{B}) \propto (1+r)^2. \tag{8.153}$$

Going back to Eq. (8.117) and accounting for contribution of the ϵ related term leads to

$$Q \propto \int \prod_m \frac{d^2 z_m}{\pi} e^{-N\left[\epsilon r + \ln(1+r) + r\frac{|\omega|^2}{1+r}\right]}, \tag{8.154}$$

where the constant of proportionality is irrelevant since the density-of-states we are after only depends on *derivatives* of the logarithm of Q.

A shortcut In this rather tedious derivation, we expressed the integral *exactly* in terms of the variables $S_1 = \sum_k z_k'^2$, $S_2 = \sum_k z_k''^2$, and $S_3 = \sum_k z_k' z_k''$. The result we obtained was valid for any N, and we indeed validated it for $N = 1$. We then found out that for large N, a significant simplification arises, and the Jacobian associated with the transformation to spherical coordinates highly favors the vectors z_k' and z_k'' to be orthogonal to each other – i.e., the term S_3 should vanish for large N for the region of subspace which dominates the integral over the variables z_k. Similarly, $S_1 = S_2$ for the saddle-point dominating the integral (in this case, in the relevant *three* dimensional space – as the integral only depended on S_1, S_2, S_3). If we would have utilized this insight a priori (based on physical intuition, to be justified *a posteriori*, at least checking for self-consistency), the calculation can be simplified significantly. In this case, one can define N complex variables as

$$t_k \equiv \sum_j H_{kj} z_j, \tag{8.155}$$

and with some straightforward algebra Eq. (8.117) can be written (for $\tau = 0$) entirely in terms of the t_k variables (under the assumptions $S_1 = S_2$ and $S_3 = 0$). The latter are Gaussian, and by separating them into their real and imaginary parts and performing the necessary Gaussian integration, Eq. (8.154) can be readily derived.

We have to evaluate the integral over the (complex) z variables. Since the expression is symmetric in the $2N$ dimensional space, we can switch to spherical coordinates, where

$$Q \propto \int_0^\infty d\rho \rho^{2N-1} e^{-N\left[\epsilon r + \ln(1+r) + r\frac{|\omega|^2}{1+r}\right]}, \tag{8.156}$$

with $r = \rho^2/N$, and we omitted the prefactor containing the volume of the $2N$ dimensional unit-sphere. We can define $\sigma \equiv \frac{1}{r}$, to find that

$$-N \log(1 + r) = -N \log(1 + \sigma) + N \log(\sigma). \tag{8.157}$$

Absorbing $e^{N \log(\sigma)} = \sigma^N$ into the prefactor, we find that

$$Q \propto \int_0^\infty \sigma^N \rho^{2N-1} e^{-N\left[\epsilon/\sigma + \log(1+\sigma) + \frac{|\omega|^2}{1+\sigma}\right]} d\rho. \tag{8.158}$$

Writing the integral over σ, using $d\rho \propto \frac{d\sigma}{\sigma^{3/2}}$, we find that

$$Q \propto \int_0^\infty \sigma^N \frac{1}{\sigma^{3/2}\sigma^{N-\frac{1}{2}}} e^{-N\left[\epsilon/\sigma+\log(1+\sigma)+\frac{|\omega|^2}{1+\sigma}\right]} d\sigma. \tag{8.159}$$

Finally,

$$e^{N\phi} \propto \int_0^\infty \frac{d\sigma}{\sigma} e^{-N\left[\epsilon/\sigma+\log(1+\sigma)+\frac{|\omega|^2}{1+\sigma}\right]}. \tag{8.160}$$

In the large N limit, we can proceed to evaluate this integral using a saddle-point approximation (see Appendix E for a brief description of this method); The integral will be dominated by the region around a particular σ_0 where $f'(\sigma_0) = 0$, the function f being the exponent:

$$f \equiv -\left[\epsilon/\sigma + \log(1+\sigma) + \frac{|\omega|^2}{1+\sigma}\right]. \tag{8.161}$$

To leading order, we then have $\phi = f(\sigma_0)$. Thus,

$$-\epsilon/\sigma_0^2 + \frac{1}{1+\sigma_0} - \frac{\omega^2}{(1+\sigma_0)^2} = 0. \tag{8.162}$$

Recalling that $\omega = x + iy$, we then have

$$G = \frac{1}{2}\left[-\frac{\partial\phi}{\partial x} + i\frac{\partial\phi}{\partial y}\right]; \nabla^2\phi = -4\pi\rho. \tag{8.163}$$

Let us try to naively set $\epsilon = 0$. We then find that

$$1 + \sigma_0 = \omega^2. \tag{8.164}$$

This would imply a *negative* σ_0 inside the unit circle – but since by definition the variable $\sigma > 0$ we cannot use this approach inside the unit circle. Inside the unit circle it is clear that for any ϵ, however small, there will be a solution to Eq. (8.162) with $\sigma > 0$. In this case we will have to keep ϵ finite, evaluate ϕ, and only then take the limit $\epsilon \to 0$.

Case I: $|\omega| > 1$
In this case we have

$$\sigma_0 = x^2 + y^2 - 1 \implies \phi = 1 + \log(x^2 + y^2), \tag{8.165}$$

$$G = \frac{-2x + i2y}{x^2 + y^2}\frac{1}{2} = \frac{1}{x + iy}. \tag{8.166}$$

Since G is analytic in this domain, the Laplacian vanishes and the DOS vanishes there. The potential scales as $\log(r^2)$: symmetrically. But this does not yet imply a constant DOS inside the circle (can you see why not?).

Case II: $|\omega| < 1$

In this case it is clear that σ_0 will be close to 0, and in Eq. (8.162) we must have $\sigma \sim \sqrt{\epsilon}$ such that the contribution of the first term will be of order unity:

$$\epsilon/\sigma_0^2 \approx 1 - \omega^2 \implies \sigma_0 \approx \sqrt{\epsilon}\frac{1}{\sqrt{1 - \omega^2}}. \tag{8.167}$$

We then find according to Eq. (8.161):

$$-\phi = \epsilon/\sigma_0 + \log(1 + \sigma_0) + \frac{x^2 + y^2}{1 + \sigma_0} \approx x^2 + y^2. \tag{8.168}$$

Therefore, its Laplacian will be constant! This is precisely the content of the circular law.

Beyond the circular law Consider the complex Ginibre ensemble corresponding to the joint distribution of Eq. (8.77). For a finite N, there will be "thermal" fluctuations (adopting the Coulomb gas interpretation), and thus we expect to find some eigenvalues outside the unit circle. What is the probability to find a "stray" eigenvalue far away from the pack? This is explored in Problem 8.8 (essentially finding the large deviation function mentioned in Chapter 6). Similarly, there is a finite probability that the largest magnitude eigenvalue is *smaller* than $\sqrt{2N}$ (the factor of 2 arising since the matrix elements are complex – can you see why?). This is studied in Problem 8.9, also using the Coulomb gas (electrostatic) interpretation. See also Majumdar and Vergassola (2009) and Lacroix-A-Chez-Toine *et al.* (2018) studying this problem both for Hermitian and non-Hermitian ensembles.

8.5 Summary

Recalling the discussion "On surprises" in Chapter 1, in this chapter we studied a phenomenon that is quite surprising given our knowledge of Poisson statistics, which sets our expectation that the level spacing between energy levels, buses, etc. will be exponentially distributed. In contrast, it was empirically found that the distribution is rather universal but takes an altogether different form – well captured by considering an ensemble of (Hermitian) random matrices. We next studied the distribution of eigenvalues of this ensemble, leading us to Wigner's semicircle law. We also derived the joint probability distribution of eigenvalues (which also manifested the "level repulsion"). Following Wigner and Dyson this allowed us to utilize an analogy to logarithmically interacting particles in a confining potential, the "Coulomb gas." Finally, we studied an ensemble of *non-Hermitian* random matrices, finding the analog of the semicircle law in the complex plane: the circular law. We also discussed the implications of this law on the stability of ecological networks within a toy-model suggested by May.

For further reading Livan *et al.* (2018) provides an excellent and very readable follow-up. Mehta (2004) is a comprehensive text dealing primarily with Hermitian matrices and going well beyond the discussion in this chapter. For a rigorous treatment of the results discussed here for Hermitian matrices (and much beyond), see Anderson *et al.* (2010).

8.6   Exercises

8.1   Wigner's Surmise for the GUE

Repeat the calculation associated with "Wigner's surmise" for the GUE, where the matrix entries are drawn from the ensemble: $p(\boldsymbol{H}) \propto e^{-tr\,H^2}$. (note that in the GUE ensemble the off-diagonal matrix entries are complex but the matrix is Hermitian).

8.2   Level Repulsion

Consider an ensemble of real, symmetric, $N \times N$ random matrices $p\,(\boldsymbol{H}) \propto e^{-tr\,H^2}$:

(a)   When ordering the eigenvalues according to their size, the level spacing, or the difference between consecutive eigenvalues, depends on where we are within the spectrum simply because the distribution of eigenvalues is non-uniform. Show analytically that if one replaces the series of eigenvalues $\lambda_1, \lambda_2, \ldots, \lambda_N$ by the series $C(\lambda_1), C(\lambda_2), \ldots, C(\lambda_N)$, where C is the cumulative distribution, then the non-uniform density-of-states is compensated for. This procedure is called unfolding.

(b)   Consider $N = 3000$, and find the eigenvalues numerically. Compare the distribution of level spacing after unfolding to that of Wigner's surmise derived in class.

(c)   Consider now the case where only a finite fraction f of the matrix elements are nonzero (but the matrix is still symmetric). Repeat (b) for $f = 0.1, 0.01$. How does the level repulsion depend on the sparsity of the matrix? What do you expect to find in the limit of small f?

(d)   Compare the distribution of eigenvalues to the semicircle law for these two cases.

8.3   Moments of the Semicircle Law

The semicircle law says that the distribution of eigenvalues x follows $p\,(x) = \frac{1}{2\pi}\sqrt{4 - x^2}$ for $x \in [-2, 2]$. In this problem, we will calculate the moments of the semicircle law.

(a)   What are the odd moments?

(b)   For the even moments I_{2n}, find a recursive relation between I_{2n} and $I_{2(n-1)}$, e.g., via trigonometric substitutions and integration by parts.

(c)   Show that the Catalan numbers $C_n = \frac{1}{n+1}\begin{pmatrix} 2n \\ n \end{pmatrix}$ solve the recursive relation.

8.4 Large Eigenvalues

Consider a real and symmetric $N \times N$ matrix where the matrix elements are drawn from a uniform distribution in the interval $[0, 1]$.

(a) Show analytically that there is an eigenvector with eigenvalue scaling approximately linearly with N, and compare with numerics.
(b) Show numerically that for large N the rest of the eigenvalues still follow the semicircle law.

8.5 Numerics for Non-Hermitian Matrices

Find numerically the eigenvalues of a real, $N \times N$ *non-Hermitian* matrix, where all elements are independently chosen from a Gaussian distribution with vanishing mean and variance 1.

(a) Show that distribution of eigenvalues is approximately uniform in the disk for large N.
(b) Find the scaling of the disk radius with N.
(c)* Define and find the level repulsion between the eigenvalues.

8.6 The Elliptic Ensemble

Consider the elliptic ensemble, which we defined as

$$p(H) \propto e^{-\frac{N}{2(1-\tau^2)} tr(HH^t - \tau H^2)}.$$

(a) Simulate the elliptical ensemble for $\tau = -0.5$ (a *negative* value).
(b) What happens to the distribution of eigenvalues as $\tau \to -1$? Prove your result for the distribution of eigenvalues in this limit. *Hint: you may want to consider the moment method.*

8.7 Other Forms of non-Hermitian Random Matrices

Consider an $N \times N$ non-Hermitian random matrix A with structure inspired by neural networks (specifically, synaptic matrices whose entries A_{ij} measure the strength of connectivity between neurons i and j). Since each neuron is either excitatory or inhibitory, the columns of A tend to be mostly positive or mostly negative. The first $N/2$ columns of A are "excitatory" and the last $N/2$ are "inhibitory." The matrix elements in excitatory (inhibitory) columns are drawn from a normal distribution with mean $\mu_E/\sqrt{N}$ ($\mu_I/\sqrt{N}$) and variance $1/N$. The means are balanced so that $\mu_E > 0$, $\mu_I < 0$ and $\mu_E + \mu_I = 0$.

(a) Find numerically the eigenvalues of A and show that there are outlier eigenvalues that fall *far* outside the unit circle. Consider, for example, $\mu_E = 5$, $\mu_I = -5$, and $N = 3000$.
(b) Show that we can write $A = J + M$ where M is a deterministic matrix and J is a random matrix with elements drawn from a normal distribution with zero mean and variance $1/N$. What is the form of M? What are the eigenvalues of M?
(c) Construct a matrix A' from A by subtracting the same constant from each element in a row so that each row of A' sums to zero; equivalently $A'\vec{u} = \vec{0}$ for $\vec{u}$

such that $u_i = 1$. Find numerically the eigenvalues of A' and show that A' does not have outlier eigenvalues like those found in (a). What is the distribution of eigenvalues inside the unit circle?

(d) Show that we can write $A' = J' + M'$, where J' is a random matrix and M' is deterministic, so that the eigenvalues of J' and $J' + M'$ are the same. Find J' and M' in terms of J and M. *Hint:* Show that the vector $\vec{u}$ with identical components $u_i = 1$ is a common eigenvector and then consider the eigenvectors of J' in the $N - 1$-dimensional subspace orthogonal to $\vec{u}$, and use them to construct the eigenvectors of $J' + M'$.

(e) Use the results of (d) to explain the eigenvalue distribution of A' found in (c). Problem credit: Po-Yi Ho.

8.8 Non-Hermitian Random Matrices: Level Repulsion and Extreme Values

Consider a complex random *non-Hermitian* matrix, with real and imaginary parts of entries drawn from a normal distribution with mean zero and variance $1/N$. As shown by Ginibre, the joint probability distribution of this matrix ensemble is

$$p(\lambda_1, \ldots, \lambda_N) \propto e^{-\frac{N}{2} \sum_i |\lambda_i|^2} \prod_{i<j} |\lambda_i - \lambda_j|^2.$$

(a) Numerically find the eigenvalues of a 1000×1000 matrix drawn from this ensemble. For each eigenvalue, compute the Euclidean distance between that eigenvalue and its closest neighbor, among the 1000 eigenvalues, in the complex plane. Repeat this procedure 2000 times, and make a histogram of the distance distribution. Separately, for 1000 randomly and uniformly distributed points in the unit disk, compute the Euclidean distance between each point and its nearest neighbor. Repeat this procedure 2000 times and make a histogram of the distance distribution as well. What do the distributions look like when compared with each other, especially for small spacings? Explain the first distribution qualitatively in light of the joint distribution of eigenvalues, $p(\lambda_1, \ldots, \lambda_N)$. *Hint:* consider fitting the left tails of both distributions to power laws.

Consider now the distribution of the largest eigenvalue (measured by its complex modulus, i.e., distance to the origin) of matrices drawn from this ensemble.

(b) Using the Coulomb gas statistical mechanics analogy, find the probability that the largest eigenvalue λ_{max} is located at a distance r away from the origin in the complex plane, assuming that r is large. You should determine both leading and subleading order terms in r. *Hint:* what is the energy of dragging a charge out to a distance r?

8.9 Extreme Value Statistics of non-Hermitian Random Matrices: *Inner* Tail*

Consider an $N \times N$ non-Hermitian random matrix where all elements are complex variables, whose real and imaginary values are independently drawn from a Gaussian distribution with variance $1/2$ and vanishing mean.

Our goal in this problem will be to find (approximately) the probability distribution that the eigenvalue with the largest magnitude is smaller than $r\sqrt{N}$, with $r < 1$. You may assume N to be large. To this end we will utilize the form of the joint probability distribution, and rely on the mapping to a Coulomb gas.

(a) Show that the charges *inside* the disk will be in mechanical equilibrium if they are uniformly distributed with a certain density, which you should determine. Does this account for all of the charge in the system? If not, where can the other charges be?

(b) Find the difference in energy, as a function of r, between the configuration determined in part (a) and the charge distribution corresponding to the circular law.

(c) What does (b) imply (approximately) for the probability of observing this configuration?

9 Percolation Theory

For our path in life . . . is stony and rugged now, and it rests with us to smooth it. We must fight our way onward. We must be brave. There are obstacles to be met, and we must meet, and crush them!'

(Charles Dickens, *David Copperfield*)

9.1 Percolation and Emergent Phenomena

Percolation is a beautiful example of how simple components may conspire to give a rather striking phenomenon – in this case, of a sharp transition. To illustrate the concept, consider an $N \times N$ grid, where in each point we plant a bush (not a tree – so as not to confuse with the graph terminology that follows later in this chapter!) with probability p. We now put the bushes on the left side of the grid on fire, and ask the question: What is the probability that the fire reached the other side of the "forest"? (i.e., that the forest fire percolated). Clearly, when p is small this probability is 0, while for $p \to 1$ we will almost certainly find a conducting path. Fig. 9.1 shows a result of a numerical simulation, where the probability was estimated for a given p by running many runs and checking which fraction of them have a conductive path.

The sharpness of the transition is striking, and perhaps unexpected. As $N \to \infty$, the transition becomes sharper and sharper, i.e., there is a critical p at which percolation occurs. From Fig. 9.1 we estimate this critical probability p_c to be close to 0.59, and indeed, more elaborate studies can pinpoint the transition to remarkable accuracy, $p_c \approx 0.59274605079210$ (Jacobsen 2015).

Let us comment on the simple code that generates Fig. 9.1, which uses a so-called depth-first-search (DFS) (see, for example, Knuth 1997). First, we initiate a lattice, and "plant" the bushes randomly according to the probability p. Next, we initiate a stack, and put the coordinates of all the bushes on the left edge of the grid in the stack. We also mark them as "visited." Now, as long as the stack is not empty, we repeat the following:

1. Take the coordinates of a bush out of the stack.
2. Check whether it has neighboring bushes in its four sides, which *have not been marked yet*. If so, insert them to the stack and mark them as visited.
3. If we reached a bush on the other side of the grid, we may stop – we have percolated.

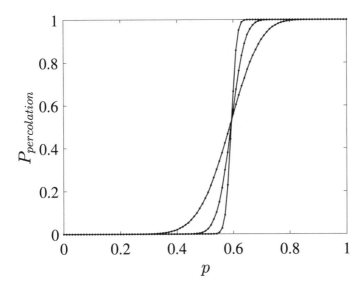

Figure 9.1 Site percolation on the square lattice. Each point is evaluated by averaging between 1000–10,000 runs on grids of varying sizes. Values of N used are 10, 30, and 100, with the plots becoming more step-like as N increases.

If the algorithm completed without reaching the other side, then no percolating path exists. Using a stack is useful here, as it allows us to "penetrate" deeper into the right (if instead we were to work with a queue, the algorithm would still work but this would be a "breadth first search").

The MATLAB code on which Fig. 9.1 is based is given here (for the code - including additional files used to generate other figures in this chapter, see https://github .com/arielamir/ThinkingProbablistically):

```
N=100;
ITERATIONS=100; p_vec=0:0.02:1;
for p_indx=1:length(p_vec)
   p=p_vec(p_indx);
   counter=0;
for iteration_number=1:ITERATIONS
   A=(rand(N)<p);
   Burnt=0; %Haven't percolated yet
   flag=zeros(N);
   %Putting the first column into the stack:
   stack_size=0;
   for j=1:N
       if A(j,1)
           stack_size=stack_size+1;
           stack(1,stack_size)=j;
           stack(2,stack_size)=1;
           flag(j,1)=1;
       end;
   end;
   while stack_size>0 & ~Burnt %while there are still bushes with
                               neighbors to check:
       coors=stack(:,stack_size);
       if coors(2)==N
```

```
                Burnt=1;
        end;
        stack_size=stack_size-1;
        if coors(1)<N
            if A(coors(1)+1, coors(2))& flag(coors(1)+1, coors(2))==0
                stack_size=stack_size+1;
                stack(1,stack_size)=coors(1)+1;
                stack(2,stack_size)=coors(2);
                flag(coors(1)+1,coors(2))=1;
            end;
        end;
        if coors(1)>1
            if A(coors(1)-1, coors(2))& flag(coors(1)-1, coors(2))==0
                stack_size=stack_size+1;
                stack(1,stack_size)=coors(1)-1;
                stack(2,stack_size)=coors(2);
                flag(coors(1)-1,coors(2))=1;
            end;
        end;
        if coors(2)<N
            if A(coors(1), coors(2)+1)& flag(coors(1), coors(2)+1)==0
                stack_size=stack_size+1;
                stack(1,stack_size)=coors(1);
                stack(2,stack_size)=coors(2)+1;
                flag(coors(1),coors(2)+1)=1;
            end;
        end;
        if coors(2)>1
            if A(coors(1), coors(2)-1)& flag(coors(1), coors(2)-1)==0
                stack_size=stack_size+1;
                stack(1,stack_size)=coors(1);
                stack(2,stack_size)=coors(2)-1;
                flag(coors(1),coors(2)-1)=1;
            end;
        end;
    end;
    if Burnt
        counter=counter+1;
    end;
end;
    percolation_prob(p_indx)=counter/ITERATIONS;
end;
plot(p_vec,percolation_prob,'.-');
```

DFS vs. BFS As mentioned, DFS is potentially more efficient as the algorithm can "go deep" and find a percolating path without finding all the sites at a given distance from the initial site (which is precisely what breadth-first-search does, obtained by replacing the stack with a queue). However, there is something inefficient about the above implementation – that can be readily fixed by replacing two of the "if" statements – can you spot it?

Fig. 9.2 shows examples of *all* bushes that were planted (in gray) and all bushes that were burnt (in white), for a value of p smaller than p_c, very close to p_c, and higher than

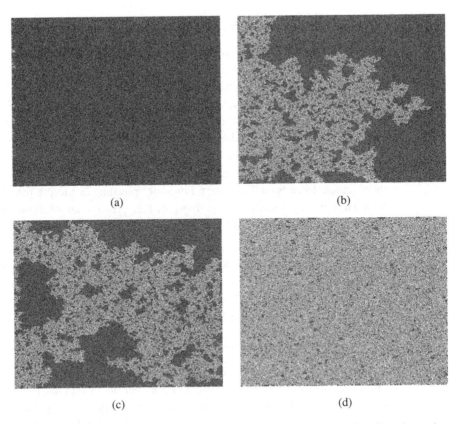

(a)

(b)

(c)

(d)

Figure 9.2 Site percolation on a 1000×1000 lattice for different values of p. Gray shows the planted bushes, white the burnt ones and black empty spaces. For (a) $p = 0.5$, (b) $p = 0.592$, (c) $p = 0.593$, and (d) $p = 0.65$. At the critical probability p_c, the network of bushes which "percolated" becomes fractal.

p_c (empty spaces are colored in black). It appears that near the percolation threshold the network of burnt bushes becomes fractal – this is indeed the case, and the fractal dimension of this network only depends on the dimensionality of the system, and not on other details, i.e., it would be the same for a honeycomb or a triangular lattice. This is referred to as "universality," and is an example of how simple interacting components may lead to rich and complex behavior – known as emergent phenomena.

Kolmogorov's zero-one law A remarkable theorem in probability theory was derived by Andrey Kolmogorov, giving conditions under which a probabilistic event will either almost surely happen or almost surely will *not* happen. Percolation on an infinite lattice, as described above, falls under the theorem's umbrella, and is a beautiful example of how rigorous probability theory (not discussed in this book!) relates to the surprising result illustrated in Fig. 9.2. For a good introduction to measure theory and probability theory, see Billingsley (2008).

9.1.1   Bond, Site, and Continuum Percolation

The above example is one of *site* percolation: Every site in the grid either includes
a bush or not. We could instead consider a model where *all* sites include bushes,
but there is a finite probability p for a bush to "infect" its neighbor. In this case
the randomness will be associated with the *bonds* (edges) rather than *sites* (vertices),
and this scenario is known as bond percolation. It can be numerically simulated in a
similar way to the one used above (e.g., using the DFS algorithm) and has the same
universal exponents as in the case of site percolation (for a given dimensionality). See
Fig. 9.3 for an example. Whether site or bond percolation should be used to model a
phenomenon depends on the context: For a fire spreading in a forest, the above model
is probably more relevant, while perhaps for the spread of an infectious disease in a
classroom we may want to use bond percolation.

Going viral We might also want to use bond percolation to describe the phenomenon
of "going viral" – though networks like the internet are not well described by a lattice,
since they are "scale-free": There is a power-law degree distribution of the nodes in
the graph, which leads to somewhat different behavior quantitatively. See also Problem
9.10 for an investigation of the "going viral" phenomenon on a real network.

Another example where bond rather than site percolation was used is in fact in the first
use of percolation theory in the context of polymer physics, in the 1940s: Paul Flory
and others modeled the gelation transition (going from liquid to a gel) as a distribution
of long polymers which are randomly cross-linked. If the density of cross-linkers is

Figure 9.3 Bond percolation on a square lattice, slightly above the percolation threshold.

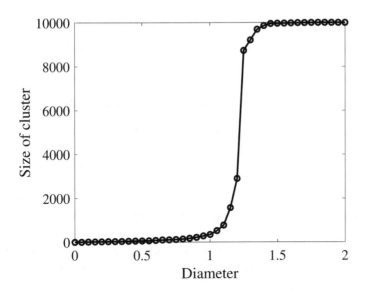

Figure 9.4 Continuum percolation of circles in 2D. The result of a single realization of the disorder with $N = 10,000$ in a 100×100 square is shown.

low, the system will be floppy (liquid phase), while at a critical density of cross-linkers, it will become rigid – when the network reaches the percolation threshold. The term "percolation" was only coined in the late 1950s. All of these developments are surprisingly recent in light of the simplicity of the models!

A different case which is very useful is that of *continuum* percolation. In this scenario there is no underlying lattice. Consider, for example, a large two-dimensional square with dimensions $L \times L$, to which we will throw small circles with radius $r \ll L$, choosing their centers randomly and uniformly in the two-dimensional square. Consider the probability of getting a path from the left side of the square to its right side which goes entirely through the circles (e.g., imagine that the circles are made of a conductive material, and we want to know whether electric currents can pass from the left of the square to its right side). Clearly, when the number of circles is small this probability is 0, while for $N \to \infty$ we will almost certainly find a conducting path. This version of percolation also shows the sharp behavior we observed in Fig. 9.1, where now percolation occurs when the filling fraction of the circles (i.e., $\pi r^2 N / L^2$) reaches a critical value (this has to be the relevant scaling, from considerations of the dimensionality). The same threshold behavior will also occur if we slightly vary the question at hand: For instance, Fig. 9.4 shows the result of a numerical simulation, where the *total number of circles* which can be reached from the LHS is numerically found for various values of the circle diameter. For each value, $N = 10,000$ circles are thrown into a square with dimensions $\sqrt{N} \times \sqrt{N}$. Note that *no* averaging is performed.

The diameter at the percolation threshold is about 1.2. This implies that the volume fraction of the circles is

$$\pi(1.2/2)^2 N / N = 1.13. \tag{9.1}$$

This is indeed very close to the known value of 1.12808737 (Mertens and Moore 2012). We will later use this result when analyzing random resistor networks.

9.1.2   Percolation Threshold for Bond Percolation on a Square Lattice

Generally, we cannot calculate the percolation threshold analytically and have to resort to numerics. However, in some cases exact solutions are possible. One simple example is that of bond percolation on a square lattice, if we *assume* that there is a sharp transition at some probability p (which we will not prove here – though some intuition for this will be attained via the renormalization group approach in the next section). This relies on the fact that the square lattice is *self-dual*: Consider Fig. 9.5, showing an $N \times (N + 1)$ grid, and consider a particular instance of bond percolation on the solid line grid from top to bottom, which corresponds to a particular *subset* of the solid edges shown (the thicker solid edges should be interpreted as the bonds which are present, while the thinner ones are only a guide to the eye). We may use it to define a particular choice of edges in the dashed line grid, by taking each dashed edge to be present if and only if the solid one cutting it is *not* present (again, in the figure only the thicker dashed edges correspond to existing bonds). According to this choice, if there is a percolating path from top to bottom in the solid line grid, it will *block* our path from left to right in the dashed line grid, and there will be *no* percolation path in

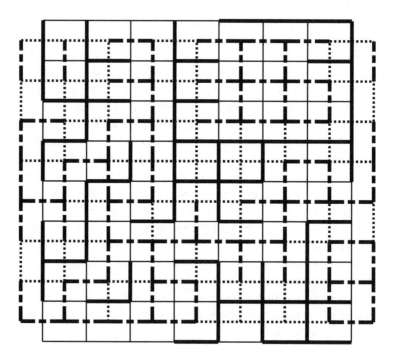

Figure 9.5 Illustration of the self-duality of an Nx$(N + 1)$ lattice, for $N = 7$.

the dashed line grid. The reverse is also true of course, and if there is a path from left to right in the dashed line grid there cannot be a path from top to bottom in the solid line grid. (Try to see which of these two scenarios occurs in Fig. 9.5!) This proves that $P_{perc}(p) = 1 - P_{perc}(1 - p)$, where P_{perc} is the probability of percolation. In particular, if we choose $p = 1/2$, we find that $P_{perc} = 1/2$, and hence it must be the percolation threshold! (this is studied numerically in Problem 9.1). Note that the fact that the grid is not perfectly square will not matter for large N.

9.1.3 Renormalization Group Approach for Site Percolation on the Square Lattice

Previously, we found numerically that in the case of site percolation on the square lattice the percolation threshold is close to 0.59. We shall now derive an approximate value for this threshold analytically, using a renormalization group (RG) approach, that will, more importantly, also give us intuition regarding the origin of the threshold behavior of the percolation transition, where the probability abruptly increases at a particular probability.

Consider a large system with dimensions $2^n \times 2^n$, and divide it into four sub-blocks. Denote the probability for each of the sub-blocks to percolate by p_n. Roughly speaking, for the entire system to percolate we need either all sub-blocks to percolate, any three of them, or the top or bottom two sub-blocks (this is not a rigorous statement though – since we may find cases where sub-blocks do not percolate from left to right but the entire system does, and vice versa). Hence, the probability for the entire system to percolate is given approximately by

$$p_{n+1} \approx f(p_n) = p_n^4 + 4p_n^3(1 - p_n) + 2p_n^2(1 - p_n)^2. \tag{9.2}$$

We can repeat this procedure again and again, each time subdividing the system into four blocks. This will lead us to an iterative map of the form $p_{n+1} = f(f(f \ldots (p_{n-m})))$. Eventually we will end up with a single site, where the probability to percolate is trivially $p_1 = p$ (with p is the probability to "plant a bush" in the lattice). Hence we may expect the probability to percolate for a large system to be well approximated by the limit of the iterative map $f(x)$ (with the initial condition given by the probability p). Fig. 9.6 shows the result of 100 iterations – it appears that the iterative map approaches zero below a certain probability, and approaches one above it – precisely as expected for percolation on a lattice based on our previous discussion.

The position of the step (approximately, the percolation threshold) can readily be found analytically, by considering the possible fates of this iterative map – in other words, the fixed points of Eq. (9.2). Clearly, $p = 0, 1$ are two solutions. But since this is a fourth-order polynomial, there are two additional solutions. One of them will be negative and irrelevant, while the second will turn out to be a good approximation for the percolation threshold: if we consider the plot of Fig. 9.6, since the iterative map approaches 0 for sufficiently small values of p and 1 for sufficiently large values, we expect there to be an intermediate value of p for which $f(p) = p$, which given the

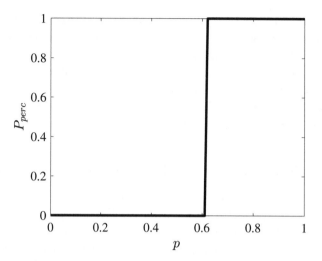

Figure 9.6 Iterating the relation of Eq. (9.2) 100 times, starting from different initial conditions (p) leads to the threshold behavior characteristic of the percolation transition.

threshold behavior of the iterative map has to coincide with the location of the step. Within the RG approach, the relation $p = f(p)$ implies that when zooming in or out of the system, the probability to percolate does not change – this is a signature of the scale-invariance of the system at the percolation threshold. At the critical point, the system does not have a well-defined length scale, and is self-similar.

Dividing Eq. (9.2) by p_n (which is assumed to be equal to p_{n+1}), we thus find the following equation for the percolation threshold:

$$p_c^3 - 2p_c + 1 = 0. \tag{9.3}$$

To find the other two roots, we can divide the polynomial by ($p_c - 1$) (since 1 is a root), finding that

$$
\begin{array}{r}
X^2 + X - 1 \\
\hline
X - 1) \; X^3 \qquad - 2X + 1 \\
\underline{- X^3 + X^2} \\
X^2 - 2X \\
\underline{- X^2 + X} \\
- X + 1 \\
\underline{X - 1} \\
0.
\end{array}
$$

Therefore, the other two roots are $\frac{-1 \pm \sqrt{5}}{2}$. One of them is negative and is hence ruled out, while the other is the inverse of the golden ratio, and is approximately equal to 0.612 – very close to the numerically established value we found earlier, despite the crude approximation made.

RG approach on a larger block The fact that the simple RG scheme described above led to a rather accurate value of the percolation threshold might hint that repeating it on, e.g., a 3×3 subdivision might lead to significantly better results. Try calculating the relevant polynomial associated with the RG in this case, and find its relevant root. You might be surprised that the results are not very different. For a related study, see Ziff and Newman (2002).

9.2   Percolation on Trees – and the Power of Recursion

Consider now a different topology of the connectivity graph, of an infinite tree, where every vertex has d "daughter" vertices. We will be considering bond percolation on this tree, where every edge is present with probability p (i.e., the probability of a given vertex being connected to its two daughters is p^2, and to none of them $(1-p)^2$). see the RHS of Fig. 9.7. This setup is very similar to the Bethe lattice (named after Hans Bethe), where every vertex has a constant degree M, as illustrated in the LHS of Fig. 9.7.

Consider the "root" of the tree, and let's ask the following question: what is the probability that the number of sites connected to the root is infinite? (corresponding to "percolation" on this network). Throughout the calculations that follow, the self-similar nature of the tree will make *recursion* a powerful tool to solve the problem.

To answer this, note that the probability of a particular sub-branch being (a) connected and (b) having an infinite lineage ("downstream" of the root) is pp_∞. Furthermore, the probability of having a finite tree is the product of the probabilities that all of the connected sub-branches have finite lineages, hence

$$(1 - p_\infty) = (1 - pp_\infty)^d. \tag{9.4}$$

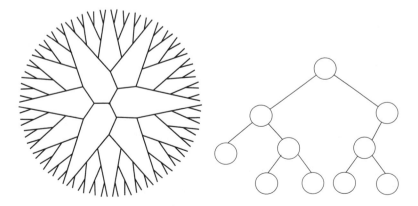

Figure 9.7 (left) A subset of the (infinite) Bethe lattice, where the degree of every site is precisely M (here $M = 3$). (right) A branching process with $d = 2$, where every vertex has either zero, one, or two "daughters."

There always exists a solution $p_\infty = 0$, but for large enough p (above the percolation threshold p_c) we will have finite solutions where p_∞ is nonzero.

Before discussing the solution of this equation in more detail, let us take an interlude that will help us find the percolation threshold. Consider the *expectation value* m of the number of sites connected to the root (which might be finite or infinite). Assuming that m is finite we have

$$m = mpd + 1, \tag{9.5}$$

hence

$$m = \frac{1}{1 - pd}. \tag{9.6}$$

This only makes sense for $pd < 1$. Hence, for $p > \frac{1}{d}$ we must have a diverging expectation value! This suggests that $p_c = 1/d$ – and hence we expect to find a nonvanishing probability p_∞ above this value. In sharp contrast to the case of percolation on a lattice, here p_∞ will not jump from 0 to 1 at the percolation threshold, but will monotonically increase from 0 to 1. To see this, consider Eq. (9.4) for $d = 2$. This is a quadratic equation, whose (nonzero) solution gives

$$p_\infty = \frac{2p - 1}{p^2}. \tag{9.7}$$

Clearly, this solution is only relevant for $p > \frac{1}{2}$, above which this predicts a positive value of p_∞, described by a concave, monotonic increase. This increase will continue until $p_\infty = 1$ when $p = 1$. As mentioned, this behavior is different from the threshold type behavior of percolation in lattices of finite dimension – such as the 2D lattice examples we saw – where the probability abruptly changes from zero to one. Also, note that it is not a priori obvious why for $p > \frac{1}{2}$ the nonvanishing solution of Eq. (9.7) is the relevant one. The fact that the average number of connected components diverges does not necessarily imply that $p_\infty \neq 0$. Nevertheless, this can be shown by considering a more subtle variant of the argument we used: Instead of looking for the fixed point solution of Eq. (9.9), consider the equation for $P_n(p)$, the probability that the path from the origin goes for at least n generations. One then finds the following recursive equation for P_n:

$$P_{n+1} = 1 - (1 - pP_n)^d, \tag{9.8}$$

where here $d = 2$. It is now easy to show that for $pd > 1$, using Eq. (9.8) iteratively leads to a nonzero solution (for instance, by considering the slope of the iterative map at zero).

Armed with the intuition gained from the exact solution to the $d = 2$ case (plotted in Fig. 9.8), we may similarly consider the case of $d > 2$. We may consider the regime where p_∞ is nonzero but small. We have from Eq. (9.4):

$$(1 - p_\infty) \approx 1 - dpp_\infty + \binom{d}{2} p_\infty^2 p^2. \tag{9.9}$$

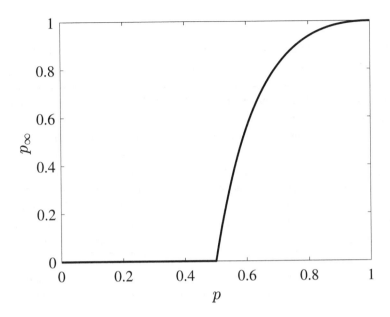

Figure 9.8 Dependence of p_∞ on p on an infinite tree for $d = 2$.

The solution is therefore

$$p_\infty \approx \frac{dp - 1}{\binom{d}{2} p^2}. \tag{9.10}$$

Identical reasoning to that applied to the $d = 2$ case shows that whenever a positive solution exists, it is the correct one. Hence also in this case the qualitative behavior is similar to Fig. 9.4. Note, however, that plugging in $p = 1$ to the formula in Eq. (9.10) does not give $p_\infty = 1$: This is because Eq. (9.10) is an approximate form only valid when p_∞ is small (except for $d = 2$, where it is exact – since no additional terms appear in the binomial expansion).

It is easy to modify these calculations to find $\tilde{p}_\infty$ and $\tilde{m}$ for the Bethe lattice, where the degree of every vertex is M (in the tree example, the degree of the root is different from that of other vertices). By the same logic, we have

$$1 - \tilde{p}_\infty = (1 - pp_\infty)^M, \tilde{m} = 1 + mpM, \tag{9.11}$$

where p_∞ and m are the quantities we calculated for the previous case (of a tree) with $d = M - 1$. Clearly, the percolation threshold will still occur at $p_c = 1/d$, and the qualitative behavior is unchanged.

9.3 Percolation Correlation Length and the Size of the Largest Cluster

Consider two randomly chosen points a distance r apart, in e.g., a scenario of site or bond percolation on a lattice. Below the percolation threshold, we only have finite clusters, and so it is plausible that for large r the probability that the two points belong to the same cluster decays exponentially:

$$p(r) \propto e^{-r/\xi}. \tag{9.12}$$

This defines a length scale ξ (the percolation correlation length), which depends on the distance from the percolation threshold, and diverges as $p \to p_c$.

Diverging lengthscale at the percolation threshold. In fact, we can get some intuition regarding the diverging lengthscale at the phase transition from our previous discussion of trees: For instance, Problem 9.3 considers the mean *depth* of the tree for $p < p_c$, and shows that it diverges as $p \to p_c$. Related to this is the result from the previous section that the expectation value of the number of nodes connected to the root diverges at p_c. This phenomenon is also intimately related to that of critical opalescence: In physical systems, a system going through a phase transition (e.g., when temperature is tuned) becomes "milky" precisely at the phase transition, since it contains all length scales and hence can scatter all wavelengths of light.

Interestingly, the same behavior holds above p_c, as long as we demand that the two points do not belong to the infinite, percolating cluster (it can be shown that only a single infinite cluster exists).

Close to the percolation threshold we have

$$\xi \propto (p - p_c)^{-\nu}, \tag{9.13}$$

where the exponent ν is *universal*, and only depends on the spatial dimension (e.g., it can be proven that in 2D $\nu = 4/3$, regardless of the geometry of the lattice or whether it is bond, site, or continuum percolation). There are many other such universal critical exponents, and there are also profound relations between them, called scaling relations.

As an example, consider the size of the largest cluster, S_N, in a finite system with $N \sim L^d$ sites. It can be shown that

$$S_N = \begin{cases} \xi^d \log(N/\xi^d), & p < p_c \\ N^{d_f/d}, & p = p_c \\ N\alpha(p). & p > p_c \end{cases} \tag{9.14}$$

Here $\alpha(p)$ is a function determining the "strength" of the largest cluster. It should vanish at p_c (since then the largest cluster size scales only sublinearly with N) and near (and above) p_c it increases as a power-law:

$$\alpha(p) \propto (p - p_c)^\beta, \tag{9.15}$$

with β another critical exponent. Note that the result in the regime $p < p_c$ is rather plausible based on our study of extreme value distributions (for instance, we obtained the logarithmic scaling for the extreme values of an exponentially decaying function, belonging to the Gumbel universality class. Here, ξ^d represents the typical cluster size, while N/ξ^d is the approximate number of roughly independent "attempts"). This is analyzed in detail using a renormalization group approach in Bazant (2000).

Consider now the system with p close to p_c, and slightly above it, and choose the system size L to be equal to ξ. On one hand, the number of sites in the largest cluster is

$$S_N \sim (p - p_c)^\beta N = (p - p_c)^\beta \xi^d. \tag{9.16}$$

On the other hand, at the scale ξ, we will not know whether we are precisely at criticality or slightly above – hence the number of sites belonging to the largest cluster is

$$S_N \sim \xi^{d_f} = (p - p_c)^{-\nu d_f}. \tag{9.17}$$

Comparing the two we find that

$$\beta - \nu d = -\nu d_f. \tag{9.18}$$

Since the dimension of the system is explicitly involved, this is known as a *hyperscaling* relation. See Stauffer and Aharony (1994) for an extended discussion.

Universality As a large system goes through the percolation phase transition (often referred to as being "near criticality"), the details of the "fine-scaled" details – site or bond percolation, triangular, or square – are lost. The aforementioned divergence of a length scale at criticality is intimately related to this phenomenon – since the length scale diverges, we lose sight of the fine details (though, as mentioned, there is nothing universal about the value of the percolation *thresholds*). The same is true of other physical systems exhibiting a phase transition, though the critical exponents would be different in general – in which case we will refer to these systems as being in a different *universality class*. To further illustrate this, look at the qualitative similarity between Figs. 9.2 and 9.9, showing simulations of the Ising model (an important model in statistical mechanics, see for example Sethna 2006) as well as experiments on the metal-insulator transition (see, for example, Imada *et al.* 1998). You should be able to get some impression of the diverging length scale and emergence of fractal structures at the phase transition.

9.4 Using Percolation Theory to Study Random Resistor Networks

In the 1960s, it was experimentally found that the conductance of many disordered materials (i.e., not perfectly crystalline, containing defects) exhibited a particular dependence on temperature T, that was empirically found to follow (Walley and Jonscher 1968):

$$\sigma(T) \propto e^{-\left(\frac{T_0}{T}\right)^{-\beta}}, \tag{9.19}$$

with T_0 a constant (with dimensions of temperature) and $\beta < 1$ a positive constant.

It was known that conduction of these materials occurred via the so-called hopping of electrons between localized states (in contrast to crystalline materials, where the wavefunctions associated with electrons are spread throughout the crystal). It was also

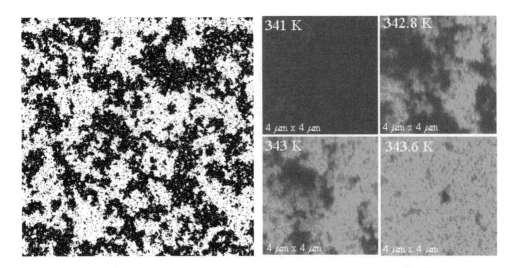

Figure 9.9 (left) Ising model near criticality (generated using https://mattbierbaum.github.io/ising.js/, with permission from Matt Bierbaum). (right) Metal-insulator transition at different temperatures. From Qazilbash *et al.* (2007). Reprinted with permission from AAAS. The color correlates with the local conductivity.

known that the rate of hopping between two states depended *exponentially* on the distance between them as well as on the energies, taking the approximate form (Miller and Abrahams 1960):

$$\sigma_{ij} \propto e^{-\frac{2r_{ij}}{\xi} - \frac{|E_i - E_j| + |E_i| + |E_j|}{2kT}}, \tag{9.20}$$

with ξ a constant (with dimensions of length), and k Boltzmann's constant, which we will set to 1 in the following (in fact, this is a simplification of the full form, and we also chose the chemical potential to be zero for convenience). Let us further assume that the positions of the localized states are chosen randomly and uniformly in a D-dimensional cube, and that the energies are chosen independently from some distribution (e.g., Gaussian) with characteristic energy scale U_0. As discussed in Miller and Abrahams (1960), an equivalent problem is the determination of the conductance (that is, the inverse of resistance) of a network of resistors, each with a value equal to $R_{ij} = \sigma_{ij}^{-1}$ (i.e., the larger the hopping rate, the smaller the value of the effective resistor between these two nodes of the network). Note that the *average* value of resistance is meaningless (think of the heavy-tailed distributions of Chapter 6!), and we will need to use more subtle reasoning.

Finding the conductance numerically For a given network, finding the conductance of the network (between two external leads) is a simple linear problem – you may consider the voltages of each of the nodes as variables, and set up a system of equations resulting from the demand that the total current into a given node must equal the total current out of it (with the value of each current proportional to the voltage difference, the prefactor being the conductance σ_{ij}). This is equivalent mathematically, but more efficient numerically, than solving Kirchhoff's equations for the currents – can you see why?

The dependence of the hopping on temperature in Eq. (9.20) is reminiscent of the Arrhenius dependence discussed at length in Chapter 4, and might lead us to believe that the conductance of the entire system should also exhibit an Arrhenius form. We will now see, using percolation theory, why this is not the case, and the model defined above in fact leads to the following dependence on temperature:

$$\sigma(T) \propto e^{-\left(\frac{T_0}{T}\right)^{-\frac{1}{D+1}}}, \tag{9.21}$$

where D is the dimensionality of the system (namely, 2 or 3).

This law was first elucidated by Sir Neville F. Mott (1969) in a simple yet non-rigorous argument (for which, among other works, he received the Nobel prize in physics). Mott's insight was to realize that it is sometimes "worthwhile" for an electron to hop from a given state to a state far away (i.e., not one of its nearest neighbor), pay the exponential penalty due to the $e^{-2r/\zeta}$ term, but *gain* overall if the state it hops to has an appropriate energy such that the second penalty of Eq. (9.20) is reduced. This could be beneficial when the temperature is sufficiently low and the energetic penalty is important. Mott's "back-of-the-envelope" calculation reproduced the temperature dependence of Eq. (9.21), which is also known as "Mott variable-range-hopping" (VRH) in light of the intuition provided above.

At the beginning of the 1970s this was put on a firmer theoretical basis via the clever use of percolation theory (Ambegaokar *et al.* 1971), which we will now outline. Note that percolation will be used here as a mathematical "tool" – nothing will physically percolate. The basic idea is as follows: Consider all the resistors in the network. Because of their exponential dependence on parameters, their values will be broadly distributed. The crux of the matter is to now consider the minimal value of resistor R_c, such that keeping all resistances with $R \le R_c$ still leads to a percolating network (i.e., to a network for which there is a continuous path from left to right). If we remove all resistors with resistance equal to or larger than R_c, by definition we are below the percolation threshold, and no current can flow through the system. Therefore, the current must flow through one or more "bottleneck resistors" with values of resistance comparable to R_c. Away from the bottleneck resistors, the current will flow through "better" resistors with $R < R_c$, but the precise value of these resistors is not important; even if they were perfect conductors, with zero resistance, the conductance of the network would nevertheless still be limited by the bottleneck resistors, and hence will be largely determined by the value of R_c. Furthermore, since the current is not forced to pass through resistors much larger than R_c, it will avoid these resistors, and they will also not be important in determining the macroscopic resistance of the network.

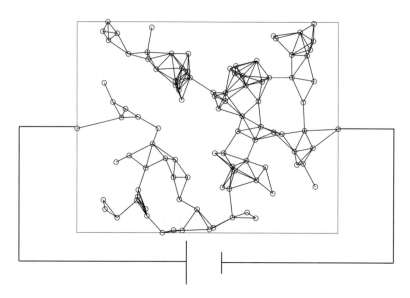

Figure 9.10 $N = 100$ points are chosen randomly and uniformly in a square with dimensions $\sqrt{100} \times \sqrt{100}$. Next, a resistor with value $e^{2r_{ij}/\zeta}$ is connected between each two points, with $\zeta = 0.05$ for this example. The conductance of the network, measured between the two leads shown, is about $2.6 \cdot 10^{-6}$. Within the percolation approach, we look for the minimal value of resistance R_c such that keeping resistors better or equal to R_c still leads to percolation between the leads. Here, $R_c = 2.3 \cdot 10^{-6}$. Can you see which resistor in the network must have precisely this value?

Therefore, R_c will be a good estimate of the resistivity of the system! This is studied and validated in Problem 9.11.

Let us demonstrate this on a particular, simpler example. Here, we only consider the spatial dependence of the resistors, i.e., $\sigma_{ij} = e^{-2r_{ij}/\zeta}$. We solve the resistor network corresponding to the (randomly) chosen positions of the states *exactly*. Second, we scan a range of conductances, and find the *minimal* value of resistance R_c such that keeping all resistors "better" than this one still leads to a connected network – this is precisely the network shown in Fig. 9.10. Remarkably, the value of R_c provides an excellent approximation to the numerically exact solution, off only by 15% in this case. Similarly, as we change the value of ζ, the value of the network resistance can span many decades, but the percolation approach still provides a good approximation to within a factor of order unity.

Now we can revisit Mott's VRH problem in full. The new problem we obtained – now of percolation theory – can be approximately solved analytically. The strategy will be as follows: We will first focus on the random energies component of the resistor network. We will restrict ourself to sites within an energy band of magnitude ΔU (for now, a free parameter), and replace the second term in Eq. (9.20) by $e^{-\Delta U/T}$ – thus getting a lower bound on the conductance (since we made all of the conductances smaller, as well as only considered a subset of them). However, by now *maximizing the conductance* over the value of ΔU, we will obtain a rather reasonable estimate of the conductivity of the system after all, that will capture the interesting scaling behavior of Eq. (9.21).

Assuming that ΔU is small compared to U_0 (which we can check once we determine the value of ΔU), the (spatial) density of the sites within the energy band we are considering is approximately

$$C \approx C_0 \frac{\Delta U}{U_0}, \tag{9.22}$$

with C_0 a constant. The larger ΔU is, the larger the energetic penalty – but the more sites we will retain.

Within the percolation approach we are utilizing, we now need to choose the minimal value of R_c such that retaining all resistors better than R_C still leads to percolation. Since the energy-related term is $e^{-\Delta U/T}$ for all the (subset) of resistors we have, we are basically retaining resistors between pairs of points i, j (in our "diluted" network of sites) such that

$$e^{-2r_{ij}/\zeta} e^{-\Delta U/T} > 1/R_c. \tag{9.23}$$

In other words, we are keeping sites within a distance d_c of each other, with

$$2d_c = \zeta[\log(R_c) - \Delta U/T]. \tag{9.24}$$

This is *precisely* a problem of continuum percolation – we know the spatial density of sites, Eq. (9.22), and we know that sites within a certain distance d_c are "connected." At the percolation threshold, we know from our previous discussion that

$$C d_c^D = P_c, \tag{9.25}$$

where C is the concentration and P_c a dimensionless constant of order unity (that does, however, depend on the system dimensionality – but not on any of the physical parameters). Therefore, we find using Eq. (9.22):

$$d_c \propto \left(\frac{U_0}{\Delta U}\right)^{1/D}. \tag{9.26}$$

This immediately leads to the following estimate for the critical conductance $\sigma_c = 1/R_c$:

$$\sigma_c \sim e^{-\Delta U/T - \frac{A}{\Delta U^{1/D}}}, \tag{9.27}$$

with A a constant independent of temperature. Optimizing now for the value of ΔU (to obtain maximal conductance) is straightforward, and leads to Mott's celebrated formula of Eq. (9.21).

A mathematically related problem is that of flow through porous media – at low Reynold's number, we can map this problem to one of a network of resistors, where the resistances correspond to the "bottlenecks" between pores in the media. The pressure at each point maps to the voltage, and liquid flow maps to the current flow (clearly, also here Kirchhoff's equations hold). For a given pressure drop, the flux through a bottleneck is known to scale as $\frac{1}{D^4}$, where D is the diameter of the bottleneck (this

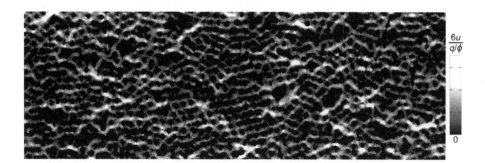

Figure 9.11 Flow through a porous network made from jamming small glass beads. The image shows that the flow is highly heterogeneous, with "hot-spots" of flow occurring at various points. Adapted Figure 1 with permission from Datta (2013). Copyright 2013 by the American Physical Society. Such behavior can be elucidated within the percolation approach of Ambegaokar *et al.* (1971).

is known as Poiseuille flow), leading to a broad distribution of effective resistances. Hence the percolation approach should be valid, and can be used to elucidate the highly heterogenous flow pattern recently observed in Datta (2013), see Fig. 9.11. Similarly, the same mathematical approach was utilized to study the behavior of random, elastic networks (Amir *et al.* 2013a).

In fact, the percolation model we outlined above is only a simplification, as it would naively suggest a "macroscopic" current would flow through one resistor. A more refined calculation elucidates that there is a lengthscale involved, much related to the percolation correlation length discussed in Section 9.3, see Shklovskii and Efros (2013) and Amir *et al.* (2013a). The stronger the disorder (i.e., the broader the distributions of the values of resistors), the larger this lengthscale. In the context of flow in porous networks, the resulting predictions have also been tested experimentally (Katz and Thompson 1986).

9.5   Summary

In this chapter we saw an intriguing example of an emergent phenomenon: Taking together very simple ingredients (nearest-neighbor interactions) we obtained a phase-transition between a regime where the physical property we were considering cannot propagate through the system (forest fire, viral spreading, etc.) to a regime where propagation is nearly certain in a large system (percolation). On tree structures (corresponding, for instance, to branching processes) we also found a phase transition albeit of a different nature, where there is no abrupt jump to a high probability of percolation but a gradual increase – yet there is a well-defined threshold below which the probability of percolation vanishes. This ubiquitous phenomenon was also a good opportunity to illustrate some useful methods and ideas in physics – such as critical exponents and the renormalization group.

For further reading Stauffer and Aharony (1994) and Sethna (2006) provide a useful physics-based introduction to the topic of percolation. Sornette (2006) and Hughes (1996) provide a more in-depth (and mathematically rigorous) discussion of most of the topics discussed in this chapter. Newman (2018) provides a comprehensive yet accessible discussion of networks (with numerous contemporary applications).

9.6 Exercises

9.1 Bond Percolation on the Square Lattice

(a) Simulate bond percolation on a 2D lattice with dimensions of $(N + 1) \times N$, for $N = 5, 10, 20$.

(b) Compute the percolation probability p_c as a function of p, the probability of having a bond between two sites. Show that all curves meet at a particular point.

(c) What is the percolation threshold for bond percolation on a square lattice?

(d) Define the width of the transition as the range to go from $p_c = 1/3$ to $p_c = 2/3$. How does the width scale with N?

9.2 Site Percolation in 3D

(a) Simulate site percolation on a simple cubic lattice and find the percolation threshold to an accuracy of 1%.

Note: we say that percolation occurs if there is a path between a given face of the cube (say, the top face) and the opposite (bottom) face. In the forest fire analogy, we light all the bushes on one face on fire and ask whether the fire reaches the opposite face.

(b) Use the renormalization-group approach to find an approximate value for the transition and compare to the results of (a).

Hint: in part (b) you may need to numerically solve for the root of a high degree polynomial.

9.3 Percolation on Bethe Lattice

Consider percolation on the Bethe lattice, a tree where each point has degree M, and the bond between neighbors on the tree exists with probability p (see Fig. 9.7).

(a) Find the percolation threshold p_c.

(b) For a probability p close to p_c, what is the average size of a connected component?

(c) For a probability p close to p_c and the specific case of $M = 3$, find the average depth of a connected component. Depth is the number of generations of connected neighbors, measured from the component's root.

(d)* For a probability p close to p_c, now for any M, find the scaling of the average depth of a connected component with p.

9.4   Continuum Percolation of Rods

Consider a square with side 1. N rods of length L and negligible width are thrown into the square, with their center uniformly chosen in the square and their angle uniformly chosen.

(a) Assuming $L \ll 1$, how does the number of rods N needed for percolation to occur from the left side to the right side of the square scale with L?

(b) Find a lower-bound for the percolation threshold by comparison with the case of percolation of circles studied in class.

(c) Find the prefactor numerically.

9.5   Application to Bacterial Growth

At each generation, a bacterium divides into two bacteria with probability p and dies with probability $1 - p$. The process begins with one bacterium.

(a) Find numerically the probability distribution $p(T)$ of the time until the colony of bacteria goes extinct.

(b) Find analytically the probability distribution $p(T)$ for p close to p_c and $T \gg 1$. *Hint*: consider the probability that the colony goes extinct *by* some time.

(c) Compare your results from (a) and (b) for $p = 0.499$.

9.6   Correlation Function on Bethe Lattice

Consider the Bethe lattice of Fig. 9.7.

(a) Calculate the correlation function $g(r)$, defined as the probability to have some connected site a distance r away (as a function of p).

(b) What happens to $g(r)$ when p is close to p_c?

9.7   Percolation on Erdős-Rényi Graphs

Consider a graph with N vertices. For every two vertices the edge between them will be present with probability p. This is an important class of random graphs known as Erdős-Rényi (ER) graphs.

(a)   Simulate an ensemble of ER graphs for $N = 10, 100, 1000$.

(b)   For each value of N find (numerically) the probability that the largest cluster is larger than $\sqrt{N}$ as a function of p, for the appropriate range of p.

(c)   What is (numerically) the approximate threshold for "percolating" on the ER graph?

(d)* Show the plausibility of the result of (c) by assuming that below the percolation threshold the finite clusters are trees, and using the results for percolation on the Bethe lattice.

9.8   Directed Percolation

Directed percolation can be viewed as adding a time dimension to usual percolation, so that, in this dimension, sites or bonds propagate in one direction only. Consider the case of directed site percolation on a square lattice, with one spatial and one temporal dimension, as depicted in Fig. 9.12. Denote as p the probability that the "daughter"

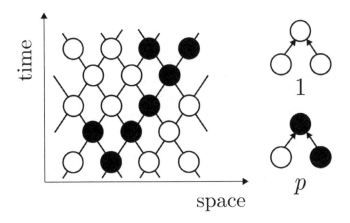

Figure 9.12 Illustration of directed percolation.

site is active given that one of the "mother" sites is active. If both mother sites are inactive, then the daughter site is also inactive with probability 1.

(a) What is the probability that the daughter site is inactive given that both mother sites are active?

(b) Denote as s_n the average fraction of active sites at a time n, and suppose that it is small. Assuming that the active sites are distributed randomly and independently at all times (i.e., neglecting spatial correlations), find a recursive relation of the form $s_{n+1} = \alpha s_n + \beta s_n^2$, where α and β are functions of p.

(c) Determine the critical probability p_c for which $p > p_c$ results in percolation. Here percolation is defined as having at least one site active at all times.

(d) The size of the connected component grows as $(p - p_c)^\gamma$ for p close to and larger than p_c. Determine the value of γ.

(e) Simulate directed percolation on the square lattice and compare your results to those found in (c), (d). You should find values of p_c and γ that deviate from the results of (c). What could be the explanation for this discrepancy?

Problem credit: Felix Wong.

9.9 Scaling of Critical Exponents

Simulate 100 runs of site percolation on square lattice with size L^2, where L varies from 32 to at least 256. Find the size of the largest cluster (using DFS for example), and test the scaling law of Eq. (9.14) (note that in two dimensions $d_f=91/48$.) Your choice of p can be for instance 0.5, 0.59274605079210, and 0.7 for the three cases.

9.10 Going Viral*

Through social network, some videos can "go viral" within a few hours. Assume that for a given video, friends share it with each other with probability p. We define "virality" as the percentage of social network users who share this video. Download the network data from www-personal.umich.edu/~mejn/netdata/hep-th.zip. This is the collaboration network of high-energy physicists.

(a) Plot the degree distribution of this network.
(b) Start with $N = 10$ people who are randomly and uniformly chosen among all the people. They all post this video online. Each bond in this network will have probability p to spread the video. Numerically find the "virality" as a function of p.

9.11 Percolation Approach to Random Resistor Networks*

Consider N points chosen randomly and uniformly in a 2D square of side $\sqrt{N}$. We put a resistor between each two points with resistance proportional to e^{Cr}, where r is the distance between the points and C is a constant. The system is connected to left and right leads located at the middle of the left and right sides of the square, as shown in Fig. 9.10 (each point is connected to both leads by resistors, the value of which is governed by the same exponential dependence).

(a) Using numerical simulations, find the ensemble average of the resistance of the system as measured between the leads for $C = 1, 2..10$ and $N = 30$.
(b) For the systems considered in (a), consider now the smallest value R of resistance such that if we remove all resistors with a value larger than R from the network, we still have a path between the two leads. Compare the values of R with those obtained in (a).

Appendix A Review of Basic Probability Concepts and Common Distributions

In this appendix we remind the reader of some basic definitions in probability theory, and properties of some distributions used in the main text. For further reading on elementary probability theory, see for example Morin (2016) or Blitzstein and Hwang (2014). Note that knowledge of measure theory is not needed for this book.

The **cumulative distribution function** (cdf) of a random variable X is given by

$$F(x) = \text{Prob}[X \leq x]. \tag{A.1}$$

F is a nondecreasing function and satisfies

$$\lim_{x \to -\infty} F(x) = 0, \tag{A.2}$$

$$\lim_{x \to \infty} F(x) = 1. \tag{A.3}$$

If X is a continuous random variable with cdf F, the probability that $a < X \leq b$ is

$$\text{Prob}[a < X \leq b] = F(b) - F(a) = \int_a^b p(x)dx, \tag{A.4}$$

where $p(x) \equiv dF(x)/dx$. p is the **probability density** (which in the main text we often refer to as the **probability distribution**, following the physicists' notation; similarly, we will refer to the cdf as the **cumulative distribution**).

Using the probability distribution, we can write expressions for the **mean** $E[X]$ and **variance** $\text{Var}[X]$ of a random variable.

$$E[X] = \int_{-\infty}^{\infty} xp(x)dx; \text{Var}[X] = E[(X - E[X])^2] = \int_{-\infty}^{\infty} (x - E[X])^2 p(x)dx. \tag{A.5}$$

A few important properties of random variables are

- **Independence**: $p(x, y) = p_X(x) \cdot p_Y(y)$ where p is a joint distribution of random variables X and Y. Note that we will often omit the subscript X, i.e., $p(X)$ will be interpreted as the probability distribution of the variable X (again, following the physics convention).
- **Conditioning**: $p_Y(y|X = x) = \frac{p(x, y)}{p_X(x)}$. If X and Y are independent, $p_Y(y|X = x) = p_Y(y)$.

- **Change of variables**: Given a continuous random variable X, define a new variable $Y = g(X)$. Then

$$p_Y(y) = \frac{p_X(g^{-1}(y))}{|g'(g^{-1}(y))|} \tag{A.6}$$

(due to the Jacobian of the transformation).

A.1   Some Important Distributions

We list below several common probability distributions, along with their mean and average, when applicable.

A.1.1   Discrete Distributions

Bernoulli distribution. A Bernoulli random variable with parameter $p \in [0, 1]$ takes the value 1 with probability p and takes the value 0 with probability $q = 1 - p$. Its mean and variance are

$$\mathbf{E}[X] = p; \mathbf{Var}[X] = p(1 - p). \tag{A.7}$$

Binomial distribution. A discrete random variable X with distribution

$$p(k) = \binom{n}{k} p^k (1 - p)^{n-k}, \qquad k \in \{0, 1, \dots, n\} \tag{A.8}$$

is said to follow the binomial distribution with parameters n (a positive integer) and $p \in [0, 1]$. Note that a Bernoulli distribution is simply a binomial distribution with $n = 1$. The sum of n independent Bernoulli random variables is a binomial random variable. It has the following mean and variance:

$$\mathbf{E}[X] = np; \mathbf{Var}[X] = np(1 - p). \tag{A.9}$$

Geometric distribution. A random variable X with the distribution

$$p(n) = (1 - p)^{n-1} p, \tag{A.10}$$

with n a positive integer, is said to be geometrically distributed. X can be interpreted as the number of Bernoulli trials needed to get one success. It has

$$\mathbf{E}[X] = \frac{1}{p}; \mathbf{Var}[X] = \frac{1 - p}{p^2}. \tag{A.11}$$

Poisson distribution. A random variable is said to have the Poisson distribution with parameter $\lambda > 0$ if its distribution function is

$$p(k) = \frac{\lambda^k}{k!} e^{-\lambda} \tag{A.12}$$

for k a nonnegative integer. The Poisson distribution describes the number of events of a Poisson process which occur in a given interval of time. It has

$$\mathbf{E}[X] = \mathbf{Var}[X] = \lambda. \tag{A.13}$$

Example: From the Binomial distribution to the Poisson distribution. Consider an event that can occur at each (small) time interval Δt with probability $p = r\Delta t$. What is the probability to have k events in a time interval T? This would be a *binomial* distribution, with $n = T/\Delta t$. Taking the limit $\Delta t \to 0$, one finds (try it!):

$$\lim_{\Delta t \to 0} p(k) = e^{-\lambda}\frac{\lambda^k}{k!}, \tag{A.14}$$

with $\lambda = rT$.

In deriving this, the following limit is useful:

$$\lim_{n \to \infty}\left(1 - \frac{\lambda}{n}\right)^n = e^{-\lambda}. \tag{A.15}$$

Also, you may find Stirling's formula helpful:

$$n! \approx \sqrt{2\pi n}\left(\frac{n}{e}\right)^n. \tag{A.16}$$

A.1.2   Continuous Distributions

Exponential distribution. A random variable is said to have the exponential distribution with parameter $\lambda > 0$ if its probability distribution is

$$p(x) = \begin{cases} \lambda e^{-\lambda x}, & x \geq 0, \\ 0, & x < 0. \end{cases} \tag{A.17}$$

The exponential distribution describes *the time between events* in a Poisson process. It has

$$\mathbf{E}[X] = \frac{1}{\lambda}; \mathbf{Var}[X] = \frac{1}{\lambda^2}. \tag{A.18}$$

Normal distribution. The (one-dimensional) normal distribution with mean $\mu \in \mathbb{R}$ and variance $\sigma^2 > 0$ has distribution function

$$p(x) = \frac{1}{\sqrt{2\pi\sigma^2}}e^{-\frac{(x-\mu)^2}{2\sigma^2}}. \tag{A.19}$$

To check that the distribution is normalized, i.e., $\int_{-\infty}^{\infty} p(x)dx = 1$, it is convenient to consider $I \equiv [\int_{-\infty}^{\infty} p(x)dx]^2$. Writing I as a two-dimensional integral we find that

$$I = \frac{1}{2\pi\sigma^2}\int_{-\infty}^{\infty}\int_{-\infty}^{\infty} e^{-\frac{(x-\mu)^2+(y-\mu)^2}{2\sigma^2}}dxdy. \tag{A.20}$$

Shifting x and y by μ and switching to polar coordinates leads to

$$I = \frac{1}{2\pi\sigma^2} \int_0^\infty \int_0^{2\pi} r e^{-\frac{r^2}{2\sigma^2}} \, d\theta \, dr. \tag{A.21}$$

This integral can be readily evaluated showing that indeed $I = 1$. It is also straight-forward to verify that σ^2 is the variance of the distribution. See Appendix B for a generalization to the multivariate case.

A.2   Central Limit Theorem

Let $\{X_k\}$ be a sequence of mutually independent random variables drawn from a common distribution. Suppose that $\mu = \mathbf{E}(X_k)$ and $\sigma^2 = \mathbf{Var}(X_k)$ exist and let $S_n = X_1 + \cdots + X_n$. Then for every fixed β,

$$\text{Prob}\left[\frac{S_n - n\mu}{\sigma\sqrt{n}} < \beta\right] \rightarrow \frac{1}{\sqrt{2\pi}} \int_{-\infty}^\beta e^{-y^2/2} dy. \tag{A.22}$$

This convergence to the normal distribution is for the case when the mean and the variance are finite. In Chapter 6 the Generalized Central Limit Theorem is discussed, which can be applied for random variables with infinite mean and/or variance.

Appendix B A Brief Linear Algebra Reminder, and Some Gaussian Integrals

In the first part of this appendix, we remind the reader of some basic linear algebra facts. For the derivations, the reader may consult with any standard textbook on linear algebra, for example Axler (2015). We shall primarily consider finite vector spaces here, which would suffice for the majority of the usage of linear algebra made in this book. The second part of this appendix builds on these results to calculate certain multivariate Gaussian integrals, used throughout the book.

B.1   Basic Linear Algebra Facts

B.1.1   Hermitian vs. non-Hermitian Matrices

The *adjoint* of a matrix A_{ij} with complex entries is the matrix:

$$A^\dagger \equiv (A^t)^*, \tag{B.1}$$

e.g., if the matrix A is *real*, its adjoint is A^t. A matrix A is said to be Hermitian if $A = A^\dagger$.

B.1.2   Eigenvalues, Eigenvectors, Characteristic Polynomials

Consider an $n \times n$ matrix A. If $|\vec{v}| > 0$ and

$$A\vec{v} = \lambda\vec{v}, \tag{B.2}$$

λ is said to be an **eigenvalue** of A, and $\vec{v}$ an **eigenvector** of A. The eigenvalues of a given matrix can be shown to be the roots of the **characteristic polynomial**:

$$det(A - \lambda I) = 0. \tag{B.3}$$

Note that since $det(A) = det(A^t)$, a matrix and its transpose share the same eigenvalues.

B.1.3   Diagonalization

A square matrix A is **diagonalizable** if it is **similar** to a diagonal matrix, which means that there exists P such that $P^{-1}AP$ is a diagonal matrix. In particular, it can be proven that

1. A Hermitian matrix is diagonalizable, and the diagonalizing matrix P can be chosen to be **unitary**, i.e.,

$$P^{-1} = (P^t)^*. \tag{B.4}$$

2. In particular, if the matrix is **real and symmetric** the diagonalizing matrix can be chosen to be **orthogonal**, i.e.,

$$P^{-1} = (P^t). \tag{B.5}$$

Furthermore, the eigenvalues of a Hermitian matrix are real, and eigenvectors corresponding to distinct (non-degenerate) eigenvalues are orthogonal to each other.

B.2   Gaussian Integrals

Gaussian integrals are needed to solve the diffusion equation in Chapter 2, and more complicated scenarios arise in Chapter 8 (and many of the problem sets throughout the book). The one-dimensional case was discussed in Appendix A. Here, we generalize it to the n-dimensional case, making good use of the linear algebra results discussed earlier in this appendix. Finally, we compute the *correlations* between two variables drawn from a multivariate normal distribution.

B.2.1   The Multivariate Normal Distribution

The multivariate normal distribution is written as

$$p(x_1, \ldots, x_n) = \frac{1}{\sqrt{(2\pi)^n det \Sigma}} \exp\left[-\frac{1}{2}\vec{x}^T A \vec{x}\right], \tag{B.6}$$

where $A = \Sigma^{-1}$, and A is a *real* and *positive-definite* matrix (i.e., possessing only positive eigenvalues) that can be chosen to be symmetric without loss of generality (can you see why?). First, let us ensure that the distribution is normalized: $\int_{\mathbb{R}^n} p(\vec{x})dx_n = 1$.

We calculate the integral by changing to a new basis where A is diagonal (we know that this is possible since the matrix is Hermitian), so that we can separate out n integrals.

$$\int e^{-\frac{1}{2}x^T A x} dx_n = \int \exp\left[-\frac{1}{2}\left(\vec{x}^T U\right)\left(U^{-1} A U\right)\left(U^{-1}\vec{x}\right)\right] dx_n$$

$$\equiv \int \exp\left[-\frac{1}{2}\vec{y}^T \widetilde{A}\vec{y}\right] |det(U)| dy_n. \tag{B.7}$$

Here $\widetilde{A}$ is a diagonal matrix, and U an orthogonal matrix. Furthermore, the Jacobian $|det(U)|$ is 1, because U is orthogonal. Therefore,

$$\int \exp\left[-\frac{1}{2}\vec{y}^T \widetilde{A}\vec{y}\right] |det(U)| dy^n = \int \exp\left[-\frac{1}{2}\sum_i y_i \widetilde{A}_{i,i} y_i\right] dy_1, \ldots, dy_n \quad \text{(B.8)}$$

$$= \prod_{i=1}^n \int e^{-\frac{y_i^2 \widetilde{A}_{i,i}}{2}} dy_i = \prod_{i=1}^n \int e^{-z_i^2} \sqrt{\frac{2}{\widetilde{A}_{i,i}}} dz_i = \sqrt{\frac{(2\pi)^n}{det\widetilde{A}}}. \quad \text{(B.9)}$$

Note that $det\widetilde{A} = det A = \frac{1}{det\Sigma}$. Putting back the factor of $\frac{1}{\sqrt{(2\pi)^2 det\Sigma}}$ in $p(\vec{x})$, we indeed find that $\int_{\mathbb{R}^n} p(\vec{x}) dx^n = 1$.

B.2.2   Correlations between Gaussian Variables

In Section 8.4.2 essentially the same strategy of the previous subsection is utilized (i.e., of diagonalizing A, performing the calculation in the new basis, and returning to the original one). This is used to compute the correlations between two variables drawn from the multivariate normal distribution of Eq. (B.6), finding that

$$\langle x_i x_j \rangle = (A^{-1})_{ij}. \quad \text{(B.10)}$$

This result is used in Chapter 8, as well as some of the problems sets.

Appendix C Contour Integration Fourier Transform Refresher

This appendix is meant to serve as a "refresher" – readers are assumed to be familiar with the notions of meromorphic functions, contour integration, etc. For a mathematically rigorous reference for complex analysis, see for example Stein and Shakarchi (2010). For a reference for the usage of complex and Fourier analysis, see for example Hassani (2013) and Arfken, Weber, and Harris (2012). For a concise summary of these techniques (yet beyond that provided in this appendix), see Mathews and Walker (1970).

C.1 Contour Integrals and the Residue Theorem

The residue theorem states that for a meromorphic function $g(z)$,

$$\int_C g(z)dz = 2\pi i \sum (\text{residue enclosed}).$$ (C.1)

We say that the function has a pole of order n at z_0 if $g(z)(z - z_0)^n$ is nonzero and analytic in a neighborhood of z_0 (for $n = 1$, this is known as a "simple pole"). The residue of an nth order pole is given by

$$Res = \frac{1}{n-1!} \lim_{z \to z_0} \frac{d^{n-1}}{dz^{n-1}} \left[(z - z_0)^n g(z)\right]_{z=z_0}.$$ (C.2)

Note that contours are evaluated in the positive direction (i.e., counterclockwise).

C.2 Fourier Transforms

Here, we briefly review Fourier series and Fourier transforms, use of which is heavily made in Chapters 5 through 7.

C.2.1 Fourier Series

Consider a continuous function $f(\theta)$ with $0 < \theta < 2\pi$. This function $f(\theta)$ can be expanded into **Fourier series** as follows

$$f(\theta) = \frac{A_0}{2} + \sum_{n=1}^{\infty} (A_n \cos n\theta + B_n \sin n\theta).$$ (C.3)

Since the cosines and sines form an orthonormal basis of the space, we have for $n, m > 0$:

$$\frac{1}{\pi} \int_0^{2\pi} \cos(m\theta)\cos(n\theta)d\theta = \delta_{m,n}, \tag{C.4}$$

$$\frac{1}{\pi} \int_0^{2\pi} \sin(m\theta)\sin(n\theta)d\theta = \delta_{m,n}, \tag{C.5}$$

$$\int_0^{2\pi} \cos(m\theta)\sin(n\theta)d\theta = 0. \tag{C.6}$$

Using this we can obtain a representation for the coefficients

$$A_n = \frac{1}{\pi} \int_0^{2\pi} f(\theta)\cos(n\theta)d\theta, \tag{C.7}$$

$$B_n = \frac{1}{\pi} \int_0^{2\pi} f(\theta)\sin(n\theta)d\theta. \tag{C.8}$$

(Can you see where the factor of $1/2$ in front of A_0 in Eq. (C.3) comes from?)

C.2.2   Complex Fourier Series

We can generalize the Fourier series to periodic functions with arbitrary period L. The function can be represented as

$$f(x) = \frac{1}{L} \sum_{n=-\infty}^{\infty} a_n e^{2\pi i n x/L}, \tag{C.9}$$

with the (complex) coefficients corresponding to integer n defined by

$$a_n = \int_0^L f(x)e^{-2\pi i n x/L}dx. \tag{C.10}$$

C.2.3   Fourier Transforms

As $L \to \infty$, we can define $y \equiv \frac{2\pi n}{L}$ and consider the continuous function $g(y) = a_n$. Applying this to the complex Fourier series from the previous subsection, we find that

$$f(x) = \frac{1}{L} \sum_{n=-\infty}^{\infty} a_n e^{2\pi i n x/L} \to \frac{1}{2\pi} \int_{-\infty}^{\infty} g(y)e^{ixy}dy. \tag{C.11}$$

Therefore,

$$f(x) = \frac{1}{2\pi} \int_{-\infty}^{\infty} g(y)e^{ixy}dy, \tag{C.12}$$

$$g(y) = \int_{-\infty}^{\infty} f(x)e^{-ixy}dx. \tag{C.13}$$

$g(y)$ is the **Fourier transform** of $f(x)$, and $f(x)$ is the **inverse Fourier transform** of $g(x)$. Note that occasionally the Fourier transform is defined with an additional prefactor of $\frac{1}{\sqrt{2\pi}}$, in which case the prefactor of the inverse Fourier transform becomes $\frac{1}{\sqrt{2\pi}}$ instead of $\frac{1}{2\pi}$. In this book we will **not** use this alternative definition.

We can combine equations C.12 and C.13 to eliminate $g(y)$.

$$f(x) = \frac{1}{2\pi} \int_{-\infty}^{\infty} dy e^{ixy} \int_{-\infty}^{\infty} f(x')e^{-ix'y}dx' \tag{C.14}$$

$$= \int_{-\infty}^{\infty} dx' f(x') \frac{1}{2\pi} \int_{-\infty}^{\infty} e^{i(x-x')y}dy \tag{C.15}$$

$$\equiv \int_{-\infty}^{\infty} dx' f(x')\delta(x - x'). \tag{C.16}$$

Here $\delta(x)$ is a **Dirac δ-function**. Thus we found an integral representation of it: $\frac{1}{2\pi} \int_{-\infty}^{\infty} e^{ixy}dy = \delta(x)$. It has the properties

$$\delta(x) = 0, \qquad\qquad x \neq 0, \tag{C.17}$$

$$\int_{-a}^{b} \delta(x)dx = 1, \qquad\qquad a, b > 0. \tag{C.18}$$

An important property of the Dirac δ-function that naturally follows is

$$\int f(y)\delta(x - y)dy = f(x). \tag{C.19}$$

See the box in Section 5.4 for further discussion and a more "refined" interpretation.

C.2.4   Example – Fourier Transform of a Gaussian in 1D

Consider a function $f(t)$ in the time domain.

$$f(t) = \frac{1}{\sqrt{2\pi\sigma_t^2}} e^{-t^2/2\sigma_t^2}. \tag{C.20}$$

By taking the Fourier transform of $f(t)$, we find $g(\omega)$ in the frequency domain.

$$g(\omega) = \int_{-\infty}^{\infty} f(x)e^{-i\omega t}dt. \tag{C.21}$$

In Chapter 2, we use a contour integral to solve precisely this integral, finding that

$$g(k) = \exp\left[-\frac{\omega^2\sigma_t^2}{2}\right]. \tag{C.22}$$

Thus, the Fourier transform of a Gaussian function is another Gaussian function. In particular, the variance of $g(k)$ is $\sigma_\omega^2 = 1/\sigma_t^2$. This implies that the narrower a function is in the time domain, the wider it is in the frequency domain. In particular, as $\sigma_t \to 0$, $\sigma_\omega \to \infty$. In other words, the Fourier transform of a δ-function is a constant (and vice versa).

Appendix D Review of Newtonian Mechanics, Basic Statistical Mechanics, and Hessians

In this appendix we review some basic results in classical mechanics, with the goal of providing the reader with some intuition for the "overdamped limit" discussed in Chapter 3. We then provide an extremely concise description of the Boltzmann distribution, which is heavily used in Chapters 3 and 4, and a reminder of some results in multivariate calculus, including Taylor expansions of multivariate functions and Hessians. For further reading on the latter, see for example Marsden and Hoffman (1993).

D.1 Basic Results in Classical Mechanics

D.1.1 The Damped Harmonic Oscillator

In Chapter 3, we mentioned that we can ignore the inertia when the system is over-damped. Here we want to review the damped harmonic oscillator and in particular find when it is possible to ignore the inertial term $m\frac{dx^2}{dt^2}$.

Imagine an object of mass m attached to the end of a spring with spring constant k. As the object moves, it experiences a drag force which is proportional to its speed with coefficient γ:

$$F = -kx - \gamma\frac{dx}{dt} = m\frac{dx^2}{dt^2}. \tag{D.1}$$

We can solve this second-order linear ODE by guessing a solution $x \propto e^{st}$ (with complex s). We find that $s = -\omega_0\left(\zeta \pm i\sqrt{1-\zeta^2}\right)$, with $\omega_0 = \sqrt{k/m}$ and $\zeta = \frac{\gamma}{2\sqrt{mk}}$. This implies that x should exhibit damped *oscillations* if $\zeta < 1$ and exponentially decay without oscillations if $\zeta > 1$. The former is referred to as the **underdamped** regime and the latter as the **overdamped** regime.

Note that if $\zeta \gg 1$, the solution for x agrees with that of the first-order ODE without the inertial term.

$$x(t) = e^{-\omega_0\zeta t}\left(c_1 e^{-\omega_0\sqrt{\zeta^2-1}t} + c_2 e^{\omega_0\sqrt{\zeta^2-1}t}\right) \tag{D.2}$$

$$\approx c_1 e^{-2\omega_0\zeta t} + c_2 e^{-\frac{\omega_0}{2\zeta}t}. \tag{D.3}$$

Since ζ is large, the first term is negligible. Thus, we have $x(t) \approx x_0 e^{-\frac{\omega_0}{2\zeta}t}$ as $\zeta \to \infty$. This solves $2\zeta\omega_0 \frac{dx}{dt} + \omega_0^2 x = 0$. This is what is meant by ignoring the inertia in the overdamped limit.

D.1.2 Terminal Velocity

In Chapter 3, it is mentioned that particles should move at constant velocity $v = \mu mg$ in the overdamped limit. We can do a similar calculation to the previous subsection to find the terminal velocity. In this case, the object is in a gravitational potential, and the equation of motion is

$$F = mg - \gamma\frac{dx}{dt} = m\frac{d^2x}{dt^2}. \tag{D.4}$$

In the overdamped limit, we may neglect the term $m\frac{d^2x}{dt^2}$, finding the terminal velocity $v_t = \frac{mg}{\gamma} = \mu mg$, where $\mu = 1/\gamma$. Note that once the terminal velocity is reached the velocity stays constant, unlike in the previous example of the damped harmonic oscillator.

D.2 The Boltzmann Distribution and the Partition Function

Consider a system with (discrete) energy levels $E_1, E_2 \ldots E_N$. At thermal equilibrium, the probability of finding the system in the jth energy level is proportional to

$$p_j \propto e^{-E_j/k_BT}, \tag{D.5}$$

with T the temperature (measured in Kelvin) and k_B the Boltzmann constant (often set to 1 in this book). The inverse of the normalization constant is defined as the *partition function Z*, i.e.,

$$Z \equiv \sum_j e^{-E_j/k_BT}. \tag{D.6}$$

These basic concepts in statistical mechanics play an instrumental role in Chapter 4 as well as parts of Chapter 8. For further reading see any textbook on statistical mechanics, for example, Sethna (2006).

D.3 Hessians

In Chapter 4, we discuss the escape-over-a-barrier problem, and the Hessian matrices of the potential energy at a saddle-point and a local minimum determine the escape rate. In this section, we go through the definition and some basic properties of the Hessian matrices.

D.3.1   Conditions for Extrema

For a differentiable function, we may find extremal points by setting the derivative
to zero. The sign of the second derivative at an extremum can be used to determine
whether its a maximum, minimum, or inflection point. This can be generalized to $\mathbb{R}^n$.

If $f : \mathbb{R}^n \rightarrow \mathbb{R}$, we can define the gradient vector of f as

$$\nabla f(x) \equiv \begin{bmatrix} \frac{\partial f(\vec{x})}{\partial x_1} \\ \frac{\partial f(\vec{x})}{\partial x_2} \\ \vdots \\ \frac{\partial f(\vec{x})}{\partial x_n} \end{bmatrix}. \tag{D.7}$$

Similarly, we can construct the **Hessian** matrix:

$$H(\vec{x}) \equiv \begin{bmatrix} \frac{\partial^2 f(\vec{x})}{\partial x_1^2} & \frac{\partial^2 f(\vec{x})}{\partial x_1 \partial x_2} & \cdots & \frac{\partial^2 f(\vec{x})}{\partial x_1 \partial x_n} \\ \frac{\partial^2 f(\vec{x})}{\partial x_2 \partial x_1} & \frac{\partial^2 f(\vec{x})}{\partial x_2^2} & \cdots & \frac{\partial^2 f(\vec{x})}{\partial x_2 \partial x_n} \\ \vdots & \vdots & & \vdots \\ \frac{\partial^2 f(\vec{x})}{\partial x_n \partial x_1} & \frac{\partial^2 f(\vec{x})}{\partial x_n \partial x_2} & \cdots & \frac{\partial^2 f(\vec{x})}{\partial x_n^2} \end{bmatrix}. \tag{D.8}$$

Note that $H(i, j) = H(j, i)$ according to Fubini's theorem. In other words, a Hessian
matrix is a real and symmetric matrix. As discussed in Appendix B, a real and symmetric matrix can be diagonalized by an orthogonal matrix: $H = U^{-1} \Lambda U$.

Taylor expanding the function f around x, we obtain

$$f(\vec{x} + \vec{\delta}) = f(\vec{x}) + \nabla f(\vec{x})^t \cdot \vec{\delta} + \frac{1}{2} \vec{\delta}^t H(\vec{x}) \vec{\delta} + O(|\vec{\delta}|^3). \tag{D.9}$$

Assuming that the gradient vanishes at $\vec{x}$, $\vec{x}$ is a local minimum (maximum) if
$\vec{\delta}^t H(\vec{x}) \vec{\delta} > (<) 0$ for any $\vec{\delta}$. Otherwise, $\vec{x}$ is a saddle-point. We can diagonalize the
Hessian and change to the variables $\vec{y} = U\vec{\delta}$, finding that

$$\vec{\delta}^T H(\vec{x}) \vec{\delta} = \vec{y}^t \Lambda \vec{y} = \sum_i \lambda_i y_i^2, \tag{D.10}$$

with λ_i the eigenvalues of H.

This leads us to a classification of the extremal points:

- If all of the eigenvalues of $H(\vec{x}_c)$ are positive, $\vec{x}_c$ is a local minimum.
- If all of the eigenvalues of $H(\vec{x}_c)$ are negative, $\vec{x}_c$ is a local maximum.
- If some of the eigenvalues of $H(\vec{x}_c)$ are positive and the others are negative, $\vec{x}_c$ is a
  saddle-point.

Appendix E Minimizing Functionals, the Divergence Theorem, and Saddle-Point Approximations

This appendix deals with various basic techniques in calculus, which are used throughout Chapter 8. All of the methods are succinctly discussed in Mathews and Walker (1970). For a more elaborate discussion, see for example Hassani (2013) and Arfken, Weber, and Harris (2012).

E.1   Functional Derivatives

A *functional* assigns a real number to a function. Functional derivatives generalize the concept of a derivative to functionals. For a functional $H[f]$ (with f a function of a single variable here), the functional derivative can be computed with a "test function" ϕ as

$$\int \frac{\delta H}{\delta f} \phi(x) dx = \lim_{\varepsilon \to 0} \frac{H[f + \varepsilon \phi] - H[f]}{\varepsilon}. \tag{E.1}$$

Importantly, to find the extremum of a functional we may equate its functional derivative to zero.

E.2   Lagrange Multipliers

Suppose we want to maximize (or minimize) a functional $I[y(x)]$ subject to the constraint that another functional $J[y(x)]$ attains a value J_0. It can be shown that the problem is equivalent to finding the extremum of the following functional:

$$\tilde{I}[y(x)] \equiv I[y(x)] - \lambda J[y(x)]. \tag{E.2}$$

The constant λ is called the **Lagrange multiplier**, and is determined from the constraint J_0 *after* finding the extremum of Eq. (E.2).

E.3   The Divergence Theorem (Gauss's Law)

Consider a vector field $\vec{D}$ in, e.g., three dimensions, and choose S to be a surface enclosing some volume R. The divergence theorem (also referred to as Gauss's theorem) states that

$$I = \int_R \nabla \cdot \vec{D} d^3 V = \int_S \vec{D} \cdot \vec{d}S. \qquad (E.3)$$

Importantly, in the case of a vector field falling off as K/r^2 (as in electrostatics, for a unit charge at the origin), the divergence of the field is proportional to a δ-function, and therefore the theorem states that the surface integral (the RHS) is simply given by $4\pi K$. From linearity (i.e., the superposition principle) we conclude that the integral of the electric field over a surface equals $4\pi K Q$, where Q is the charge enclosed by the surface. This result is utilized several times in Chapter 8, in the context of "Dyson's Coulomb gas" (where we have $K = 1$). Note that in that case the electric field "lives" in 2D and falls off as $1/r$: This is precisely the electric field of a uniformly charged *infinite wire*, and therefore the aforementioned results from 3D electrostatics can still be used.

For more details and examples, see Schey and Schey (2005).

E.4   Saddle-Point Approximations

Consider an integral of the form

$$I = \int dx f(x) e^{A g(x)}. \qquad (E.4)$$

Assume for simplicity that the function $g(x)$ has a single maximum at x_0. For sufficiently large A the integral will be dominated by the vicinity of x_0. By Taylor expanding $g(x)$ near the peak, we deduce that the width of the region effectively contributing to the integral scales as $O(1/\sqrt{A})$. Substituting $x = x_0 + y/\sqrt{A}$ and expanding in powers of y one finds that the integral is approximated by

$$I \approx f(x_0) e^{A g(x_0)} \sqrt{\frac{2\pi}{-A g''(x_0)}}. \qquad (E.5)$$

This is typically referred to as *Laplace's method* or as the method of *steepest descent*. A similar approach can be taken in the case where the exponent is purely imaginary, where this is known as the method of *stationary phase*. The more general case, where a contour integral is deformed to pass through a saddle-point, lends the method its general name, see Mathews and Walker (1970).

Appendix F Notation, Notation ...

This appendix summarizes the notation and terminology used in the notes, chapter by chapter. Throughout the notes we follow *physics* rather than *mathematics* notation and terminology and in particular:

- **boldface** is used to denote matrices, and $\vec{v}$ to denote vectors.
- When ambiguity arises, $P_f(x)$ will denote the probability distribution associated with the random variable f.
- The cumulative distribution will often be denoted by upper case letters, e.g., $G(x)$ or $C(x)$.
- Discrete probability distributions will often be written with a subscript, e.g., the first digit distribution is denoted p_d rather than $p(d)$.
- Often, noise terms are denoted by ξ.
- The divergence of a vector field $\vec{E}$ is denoted by $\nabla \cdot \vec{E}$. The gradient of a scalar (multivariate) function p will be denoted by ∇p.
- We will often use these acronyms: i.i.d. (independent and identically distributed); RHS (right hand side); LHS (left hand side); C.C. (complex conjugate); ODE (ordinary differential equation); PDE (partial differential equation).
- Throughout, we will use the Big-O notation $O(x)$ to denote terms of the same order as x and little-o notation $o(x)$ to denote terms of lower order.
- Often, $\equiv$ is used when defining a new variable, e.g., $x \equiv y + 3$ introduces a new variable x in terms of the previously defined variable y.
- Throughout the text, log refers to the natural logarithm (to base e), unless otherwise specified.

Chapter 1

Here, and throughout this book, we will follow the physicists' terminology of referring to a probability density function (pdf) as a "probability distribution," and referring to a "cumulative distribution" for the cumulative distribution function (cdf). Moreover, for a real random variable X we will denote the probability distribution by $p(x)$, rather than the notation f_X often used in mathematics.

Chapter 2

- F.T. denotes Fourier transform.
- $\hat{p}$ denotes the Fourier transform of the probability distribution with regards to the spatial variables.
- P is used to denote the transition matrix of a Markov chain.

Chapter 3

- ξ is used to denote the noise in the Langevin equation.
- j is used to denote the particle (or, equivalently, probability) flux, and n the particle (or probability) density.
- The notation for the size control problem is summarized in Table 3.1.
- The notation for the Black–Scholes problem is summarized in Table 3.2.

Chapter 4

- k is the Boltzmann constant, and T the (absolute) temperature.
- Γ is the escape rate.
- For the main result of the chapter (Langer's formula), η_i denote the coordinates, and $p(\vec{\eta}, t)$ is the probability distribution.
- M_{ij} is the mobility matrix.
- H^m and H^s are the *Hessian* matrices of the potential around the minimum and saddle, respectively.

Chapter 5

- $P(\omega)$ denotes the power spectrum.
- $C(t - t')$ denotes the autocorrelation function.
- $\langle\rangle$ denotes ensemble-averaging.

Chapter 6

- $\varphi(\omega)$ is often used to denote the characteristic function (often with a clarifying subscript, such as the number of elements in the relevant sum or a symbol corresponding to the random variable).
- $p_n(x)$ refers to the probability distribution of the sum of n variables.
- Cumulative distribution functions are typically denoted by $C(x)$ or $G(x)$ (the latter notation is used for the limiting distributions).

Chapter 7

- $p(x,t)$ is the probability distribution of the CTRW.
- $p(k,s)$ denotes Fourier transforming $p(x,t)$ with respect to x and Laplace transforming with respect to t.
- $\psi(t)$ is the distribution of trapping times.
- $\psi(s)$ is the Laplace transform of the trapping time distribution.

Chapter 8

- Throughout the chapter, we typically denote the random matrix drawn from the various ensembles by H.
- Determinants are denoted by $det(X)$. Traces are denoted by tr.
- When matrices are diagonalized, D denotes the diagonalized matrix, and U denotes the orthogonal/unitary transformation matrix.
- In the derivation of the circular law, x' and x'' are often used to denote the real and imaginary parts of a variable x.

Chapter 9

- p_c is the percolation threshold
- For the tree (or Bethe lattice) p_∞ is the probability to reach infinity.
- ξ is the percolation correlation length.

References

Addario-Berry, L. and Reed, B. A., Ballot theorems, old and new, in: *Horizons of Combinatorics* (pp. 9–35), Springer, Berlin, Heidelberg (2008).

Ambegaokar, V., Halperin, B. I., and Langer, J. S., Hopping conductivity in disordered systems, *Physical Review B*, 4, 8, 2612 (1971).

Ambegaokar, V., Halperin, B. I. Nelson, D. R., and Siggia, E. D., Dynamics of superfluid films, *Physical Review B*, 21, 5, 1806 (1980).

Amir, A., Cell size regulation in bacteria, *Physical Review Letters* 112, 20, 208102 (2014).

Amir, A., An elementary renormalization-group approach to the Generalized Central Limit Theorem and Extreme Value Distributions, *Journal of Statistical Mechanics: Theory and Experiment*, 1, 013214 (2020).

Amir, A. and Balaban, N. Q., Learning from noise: how observing stochasticity may aid microbiology, *Trends in Microbiology*, 26, 4, 376 (2018).

Amir, A., Hatano, N., and Nelson, D. R., Non-Hermitian localization in biological networks, *Physical Review E*, 93, 4, 042310 (2016a).

Amir, A., Krich, J. J., Vitelli, V., Oreg, Y., and Imry, Y., Emergent percolation length and localization in random elastic networks, *Physical Review X*, 3, 2, 021017 (2013a).

Amir, A., Lahini, Y., and Perets, H. B. Classical diffusion of a quantum particle in a noisy environment, *Physical Review E*, 79(5), p.050105 (2009).

Amir, A., Lemeshko, M. and Tokieda, T., Surprises in numerical expressions of physical constants, *The American Mathematical Monthly*, 123, 6, 609 (2016b).

Amir, A., Oreg, Y., and Imry, Y., $1/f$ noise and slow relaxations in glasses, *Annalen der Physik*, 18, 12, 836 (2009).

Amir, A., Oreg, Y., and Imry, Y., Localization, anomalous diffusion, and slow relaxations: A random distance matrix approach, *Physical Review Letters*, 105, 7, 070601 (2010).

Amir, A., Oreg, Y., and Imry, Y., On relaxations and aging of various glasses, *Proceedings of the National Academy of Sciences*, 109, 6, 1850 (2012).

Amir, A., Paulose, J., and Nelson, D. R., Theory of interacting dislocations on cylinders, *Physical Review E*, 87, 4, 042314 (2013a).

Anderson, G. W., Guionnet, A., and Zeitouni, O., *An Introduction to Random Matrices*, Cambridge University Press (2010).

Anderson, P. W., Absence of diffusion in certain random lattices, *Physical Review*, 109, 5, 1492 (1958).

Anderson, P. W., More is different, *Science*, 177, 4047, 393 (1972).

Arfken, G. B., Weber, H. J., and Harris, F. E., *Mathematical Methods for Physicists, A Comprehensive Guide*, 7th edn, New York: Academic (2012).

Axler, S. J., *Linear Algebra Done Right*, New York: Springer (2015).

Bachelier, L., Theory of speculation (1900) in: P. Cootner, ed., *The Random Character of Stock Market Prices*, MIT Press (1964).

Baik, J., Borodin, A., Deift, P., and Suidan, T., A model for the bus system in Cuernavaca (Mexico), *Journal of Physics A: Mathematical and General*, 39, 28, 8965 (2006).

Bardou, F., Bouchaud, J.P., Aspect, A. and Cohen-Tannoudji, C. *Lévy Statistics and Laser Cooling: How Rare Events Bring Atoms to Rest*, Cambridge University Press (2002).

Battersby, S., Statistics hint at fraud in Iranian election, *New Scientist* 202, 2714, 10 (2009).

Bazant, M. Z., Largest cluster in subcritical percolation, *Physical Review E*, 62, 2, 1660 (2000).

Bertin, E. and Györgyi, G., Renormalization flow in extreme value statistics, *Journal of Statistical Mechanics: Theory and Experiment*, 2010, 08, 08022 (2010).

Bertrand, J. *Calcul des probabilités*. Gauthier-Villars (1907).

Billingsley, P., *Probability and Measure*. John Wiley and Sons (2008).

Black, F. and Scholes, M., The pricing of options and corporate liabilities, *Journal of Political Economy* 81, 637 (1973).

Blitzstein, J. K. and Hwang, J., *Introduction to Probability*, Chapman and Hall/CRC (2014).

Bohigas, O., Giannoni, M. J., and Schmit, C., Characterization of chaotic quantum spectra and universality of level fluctuation laws, *Physical Review Letters*, 52, 1, 1 (1984).

Bressloff, P. C., *Stochastic Processes in Cell Biology*, Berlin: Springer (2014).

Brown, R. XXVII. A brief account of microscopical observations made in the months of June, July and August 1827, on the particles contained in the pollen of plants; and on the general existence of active molecules in organic and inorganic bodies, *The Philosophical Magazine, or Annals of Chemistry, Mathematics, Astronomy, Natural History and General Science*, 4, 21, 161 (1828).

Bruinsma, R., Halperin, B. I., and Zippelius, A., Motion of defects and stress relaxation in two-dimensional crystals, *Physical Review B*, 25, 2, 579 (1982).

Calvo, I., Cuchí, J. C., Esteve, J. G., and Falceto, F., Generalized central limit theorem and renormalization group, *Journal of Statistical Physics*, 141, 3, 409 (2010).

Calvo, I., Cuchí, J. C., Esteve, J. G., and Falceto, F., Extreme-value distributions and renormalization group, *Physical Review E*, 86, 4, 041109 (2012).

Chen, K., Ellenbroek, W. G., Zhang, Z., Chen, D. T., Yunker, P. J., Henkes, S., Brito, C., Dauchot, O., Van Saarloos, W., Liu, A. J. and Yodh, A. G., 2010. Low-frequency vibrations of soft colloidal glasses. *Physical Review Letters*, 105(2), p.025501.

Chechkin, A. V., Klafter, J., Gonchar, V. Y., Metzler, R., and Tanatarov, L. V., Bifurcation, bimodality, and finite variance in confined Lévy flights. *Physical Review E*, 67(1), p.010102 (2003).

Clauset, A., Kogan, M., and Redner, S., Safe leads and lead changes in competitive team sports, *Physical Review E*, 91(6), p.062815 (2015).

Coffey, W. T., Garanin, D. A., and McCarthy, D. J., Crossover formulas in the Kramers theory of thermally activated escape rates—application to spin systems, *Advances in Chemical Physics*, 117, 483 (2001).

Coffey, W. T., Kalmykov, Y. P., and Waldron, J. T., The Langevin Equation: with applications to stochastic problems in physics, *Chemistry and Electrical Engineering*. World Scientific (2004).

Datta, S. *et al.*, Spatial fluctuations of fluid velocities in flow through a three-dimensional porous medium, *Physical Review Letters*, 111, 6, 064501 (2013).

Davidsen, J. and Schuster, H. G., Simple model for $1/f^{\alpha}$ noise, *Physical Review E*, 65, 2, 026120 (2002).

Downey, A. B., The structural cause of file size distributions, in *MASCOTS 2001, Proceedings Ninth International Symposium on Modeling, Analysis and Simulation of Computer and Telecommunication Systems, IEEE*, pp. 361–370 (2001)

Doyle, P. G. and Snell, J. L., *Random Walks and Electric Networks*, Mathematical Association of America (1984).

Dutta, P. and Horn, P. M., Low-frequency fluctuations in solids: $1/f$ noise, *Reviews of Modern Physics*, 53, 3, 497 (1981).

Dyson, F. J., A Brownian-motion model for the eigenvalues of a random matrix, *Journal of Mathematical Physics*, 3, 6, 1191(1962).

Ermann, L., Frahm, K. M., and Shepelyansky, D. L., Google matrix analysis of directed networks, *Reviews of Modern Physics*, 87, 4, 1261 (2015).

Feller, W., *An Introduction to Probability Theory and Its Applications* (Vol. 1), John Wiley and Sons (2008).

Feller, W., *An Introduction to Probability Theory and Its Applications* (Vol. 2), John Wiley and Sons (2008).

Filiasi, M., Livan, G., Marsili, M., Peressi, M., Vesselli, E. and Zarinelli, E., On the concentration of large deviations for fat tailed distributions, with application to financial data, *Journal of Statistical Mechanics: Theory and Experiment*, 9, P09030 (2014).

Fisher, R. A. and Tippett, L. H. C., Limiting forms of the frequency distribution of the largest or smallest member of a sample, in: *Mathematical Proceedings of the Cambridge Philosophical Society* (Vol. 24, No. 2, pp. 180–190). Cambridge University Press (1928).

Fortin, J. Y. and Clusel, M., Applications of extreme value statistics in physics, *Journal of Physics A: Mathematical and Theoretical*, 48, 18, 183001 (2015).

Gardiner, C., *Stochastic Methods*, Berlin: Springer (2009).

Ghosh, A., Chikkadi, V. K., Schall, P., Kurchan, J., and Bonn, D., 2010. Density of states of colloidal glasses. *Physical Review Letters*, 104(24), p.248305.

Gillespie, D. T., The mathematics of Brownian motion and Johnson noise, *American Journal of Physics*, 64, 3, 225 (1996).

Ginibre, J., Statistical ensembles of complex, quaternion, and real matrices, *Journal of Mathematical Physics* 6, 3, 440 (1965).

Golding, I. and Cox, E. C., Physical nature of bacterial cytoplasm, *Physical Review Letters*, 96, 9, 098102 (2006).

Györgyi, G., Moloney, N. R., Ozogány, K., and Rácz, Z., Finite-size scaling in extreme statistics, *Physical Review Letters*, 100, 21, 210601 (2008).

Györgyi, G., Moloney, N. R., Ozogány, K., Rácz, Z., and Droz, M., Renormalization-group theory for finite-size scaling in extreme statistics, *Physical Review E*, 81, 4, 041135 (2010).

Hardy, G. H., Mendelian proportions in a mixed population, *Science*, 28, 706, 49 (1908).

Hassani, S., *Mathematical Physics: A Modern Introduction to Its Foundations*, Springer Science and Business Media (2013).

Hazut, N., Medalion, S., Kessler, D. A., and Barkai, E., Fractional Edgeworth expansion: corrections to the Gaussian-Lévy central-limit theorem, *Physical Review E*, 91, 5, 052124 (2015).

He, Y. *et al.*, Random time-scale invariant diffusion and transport coefficients, *Physical Review Letters*, 101, 5, 058101 (2008).

Ho, P. Y. and Amir A., Simultaneous regulation of cell size and chromosome replication in bacteria, *Frontiers in Microbiology* 6, 662 (2015).

Ho, P. Y., Lin, J., and Amir, A., Modeling cell size regulation: From single-cell-level statistics to molecular mechanisms and population-level effects, *Annual Review of Biophysics*, 47, 251 (2018).

Honerkamp, J., *Stochastic Dynamical Systems: Concepts, Numerical Methods, Data Analysis*. John Wiley and Sons (1993).

Hughes, B. D., *Random Walks and Random Environments*, Clarendon Press, Oxford (1996).

Imada, M., Fujimori, A., and Tokura, Y., Metal-insulator transitions, *Reviews of Modern Physics*, 70, 4, 1039 (1998).

Indyk, P., Stable distributions, pseudorandom generators, embeddings and data stream computation, in: *Proceedings of the 41st Annual Symposium on Foundations of Computer Science*, (pp. 189–197), IEEE (2000).

Jacobsen, J. L., Critical points of Potts and O(N) models from eigenvalue identities in periodic Temperley–Lieb algebras, *Journal of Physics A: Mathematical and Theoretical*, 48, 45, 454003 (2015).

Jehl, X., Sanquer, M., Calemczuk, R., and Mailly, D., Detection of doubled shot noise in short normal-metal/superconductor junctions, *Nature*, 405, 6782, 50 (2000).

Jona-Lasinio, G., Renormalization group and probability theory, *Physics Reports*, 352, 4–6, 439 (2001).

Katz, A. J. and Thompson, A. H., Quantitative prediction of permeability in porous rock. *Physical Review B*, 34, 11, 8179 (1986).

Keener, J. P., The Perron–Frobenius theorem and the ranking of football teams, *SIAM Review*, 35, 1, 80 (1993).

Klafter, J. and Sokolov, I. M., *First Steps in Random Walks: From Tools to Applications*, Oxford University Press (2011).

Knuth, D. E., *The Art of Computer Programming* (Vol. 1), Pearson Education (1997).

Knuth, D. E., *The Art of Computer Programming, Vol. II, Seminumerical Algorithms*, Addison Wesley, (1998).

Kohlrausch, R., *Annals of Physics and Chemistry* (Poggendorff) 91, 179 (1854).

Kostinski, S. and Amir, A., An elementary derivation of first and last return times of 1D random walks, *American Journal of Physics*, 84, 1, 57 (2016).

Koppes, L. J. *et al.*, Correlation between size and age at different events in the cell division cycle of Escherichia coli, *Journal of Bacteriology*, 143, 3, 1241 (1980).

Krapivsky, P. L., Redner, S., and Ben-Naim, E., *A Kinetic View of Statistical Physics*. Cambridge University Press (2010).

Krbálek, M. and Seba, P., The statistical properties of the city transport in Cuernavaca (Mexico) and random matrix ensembles, *Journal of Physics A: Mathematical and General*, 33, 26, 229 (2000).

Kriecherbauer, T. and Krug, J., A pedestrian's view on interacting particle systems, KPZ universality and random matrices, *Journal of Physics A: Mathematical and Theoretical*, 43, 40, 403001 (2010).

Lacroix-A-Chez-Toine, B., Grabsch, A., Majumdar, S. N., and Schehr, G., Extremes of 2d Coulomb gas: universal intermediate deviation regime, *Journal of Statistical Mechanics: Theory and Experiment*, 1, 013203 (2018).

Lam, H., Blanchet, J., Burch, D., and Bazant, M. Z., Corrections to the central limit theorem for heavy-tailed probability densities, *Journal of Theoretical Probability*, 24, 4, 895 (2011).

Langer, J. S., Statistical theory of the decay of metastable states, *Annals of Physics*, 54, 2, 258 (1969).

Le, J. L., Bazant, Z. P., and Bazant, M. Z., Lifetime of high-k gate dielectrics and analogy with strength of quasibrittle structures, *Journal of Applied Physics*, 106, 10, 104119 (2009).

Le, J. L., Bazant, Z. P., and Bazant, M. Z., Unified nano-mechanics based probabilistic theory of quasibrittle and brittle structures: I. Strength, static crack growth, lifetime and scaling, *Journal of the Mechanics and Physics of Solids*, 59, 7, 1291 (2011).

Lehmann, N. and Sommers, H. J., Eigenvalue statistics of random real matrices, *Physical Review Letters*, 67, 8, 941 (1991).

Livan, G., Novaes, M., and Vivo, P., *Introduction to Random Matrices*, Springer (2018).

Majumdar, S. N. and Nechaev, S., Exact asymptotic results for the Bernoulli matching model of sequence alignment, *Physical Review E*, 72, 2, 020901 (2005).

Majumdar, S. N. and Schehr, G., Top eigenvalue of a random matrix: large deviations and third order phase transition, *Journal of Statistical Mechanics: Theory and Experiment*, 1, 01012 (2014).

Majumdar, S. N. and Vergassola, M., Large deviations of the maximum eigenvalue for Wishart and Gaussian random matrices, *Physical Review Letters*, 102, 6, 060601 (2009).

Manzato, C., Shekhawat, A., Nukala, P. K., Alava, M. J., Sethna, J. P., and Zapperi, S., Fracture strength of disordered media: Universality, interactions, and tail asymptotics, *Physical Review Letters*, 108, 6, 065504 (2012).

Marsden, J. E. and Hoffman, M. J., *Elementary Classical Analysis*, Macmillan (1993).

Matan, K., Williams, R. B., Witten, T.A. and Nagel, S. R., Crumpling a thin sheet, *Physical Review Letters*, 88, 7, 076101 (2002).

Mathews, J. and Walker, R. L., *Mathematical Methods of Physics*, New York: WA Benjamin (1970).

May, R. M., Will a large complex system be stable?, *Nature* 238, 413 (1972).

Mehta, M. L., *Random Matrices*, Elsevier (2004).

Mertens, S. and Moore, C., Continuum percolation thresholds in two dimensions, *Physical Review E*, 86, 6, 061109 (2012).

Merton, R. C., Theory of rational option pricing. *The Bell Journal of Economics and Management Science*, 4, 1, 141 (1973).

Metz, F. L., Neri, I., and Rogers, T., Spectral theory of sparse non-Hermitian random matrices, *Journal of Physics A: Mathematical and Theoretical*, 52, 43, 434003 (2019).

Miller, A. and Abrahams, E., Impurity conduction at low concentrations, *Physical Review*, 120, 3, 745 (1960).

Mlodinow, L., *The Drunkard's Walk: How Randomness Rules Our Lives*, Vintage (2009).

Möbius, W., Neher, R. A., and Gerland, U., Kinetic accessibility of buried DNA sites in nucleosomes, *Physical Review Letters*, 97, 20, 208102 (2006).

Montroll, E. W. and Shlesinger, M. F., On the wedding of certain dynamical processes in disordered complex materials to the theory of stable (Levy) distribution functions, in: *The Mathematics and Physics of Disordered Media: Percolation, Random Walk, Modeling, and Simulation*, pp. 109–137, Springer, Berlin, Heidelberg (1983).

Morin, D. J., *Probability: For the Enthusiastic Beginner*, Createspace Independent Publishing Platform (2016).

Mott, N. F., Conduction in non-crystalline materials: III. Localized states in a pseudogap and near extremities of conduction and valence bands, *Philosophical Magazine*, 19, 160, 835 (1969).

Muskhelishvili, N. I. and Radok, J. R. M., *Singular Integral Equations: Boundary Problems of Function Theory and Their Application to Mathematical Physics*, Courier Corporation (2008).

Nesbitt, J. R. and Hebard, A. F., Time-dependent glassy behavior of interface states in Al- Al O_x- Al tunnel junctions, *Physical Review B*, 75, 19, 195441 (2007).

Newman, M., *Networks: An Introduction*, 2nd edn, Oxford University Press (2018).

Orlyanchik, V. and Ovadyahu, Z., Stress aging in the electron glass, *Physical Review Letters*, 92, 6, 066801 (2004).

Orr, H. A., The distribution of fitness effects among beneficial mutations. *Genetics*, 163(4), pp. 1519–1526 (2003).

Page, L. *et al.*, *The PageRank Citation Ranking: Bringing Order to the Web*, Technical Report, Stanford InfoLab, Stanford, California (1999).

Paul, W. and Baschnagel, J., *Stochastic Processes*, Springer (2013).

Pearle, P., Collett, B., Bart, K., Bilderback, D., Newman, D., and Samuels, S., What Brown saw and you can too, *American Journal of Physics*, 78, 12, 1278 (2010).

Pearson, K., The problem of the random walk, *Nature*, 72, 1865, 294 (1905).

Perrin, J., *Brownian Movement and Molecular Reality*, Courier Corporation (2013).

Pinski, G. and Narin, F., Citation influence for journal aggregates of scientific publications: Theory, with application to the literature of physics, *Information Processing and Management*, 12, 5, 297 (1976).

Porat, B., *Digital Processing of Random Signals: Theory and Methods.* Courier Dover Publications (2008).

Pugatch, R. *et al.*, Anomalous symmetry breaking in classical two-dimensional diffusion of coherent atoms, *Physical Review A*, 89, 3, 033807 (2014).

Qazilbash, M. M. *et al.*, Mott transition in VO_2 revealed by infrared spectroscopy and nano-imaging, *Science*, 318, 5857,1750 (2007).

Redner, S., *A Guide to First-Passage Processes.* Cambridge University Press (2001).

Reznikov, M., De Picciotto, R., Griffiths, T. G., Heiblum, M., and Umansky, V., Observation of quasiparticles with one-fifth of an electron's charge, *Nature*, 399, 6733, 238 (1999).

Robert, L. *et al.*, Division in *Escherichia coli* is triggered by a size-sensing rather than a timing mechanism, *BMC biology* 12, 1, 17 (2014).

Roos, M., Böcking, D., Gyimah, K. O., Kucerova, G., Bansmann, J., Biskupek, J., Kaiser, U., Hüsing, N., and Behm, R. J., Nanostructured, mesoporous Au/TiO$_2$ model catalysts–structure, stability and catalytic properties, *Beilstein Journal of Nanotechnology*, 2, 1, 593 (2011).

Rosenzweig, N. and Porter, C. E., Repulsion of energy levels in complex atomic spectra, *Physical Review*, 120, 5, 1698 (1960).

Ruderman, D. L. and Bialek, W., Statistics of natural images: Scaling in the woods, in: *Advances in Neural Information Processing Systems*, pp. 551–558 (1994).

Scher, H. and Montroll, E. W., Anomalous transit-time dispersion in amorphous solids, *Physical Review B*, 12, 6, 2455 (1975).

Schey, H. M. and Schey, H. M., *Div, Grad, Curl, and All That: An Informal Text on Vector Calculus*, New York: WW Norton (2005).

Schuss, Z., *Theory and Applications of Stochastic Processes: An Analytical Approach*, Springer Science and Business Media (2009).

Sethna, J., *Statistical Mechanics: Entropy, Order Parameters, and Complexity.* Oxford University Press (2006).

Shklovskii, B. I. and Efros, A. L., *Electronic Properties of Doped Semiconductors*, Springer Science and Business Media (2013).

Shockley, W., On the statistics of individual variations of productivity in research laboratories, *Proceedings of the IRE*, 45, 3, 279 (1957).

Soifer, I., Robert L., and Amir A., Single-cell analysis of growth in budding yeast and bacteria reveals a common size regulation strategy, *Current Biology*, 26, 3, 356 (2016).

Sommers, H. J. *et al.*, Spectrum of large random asymmetric matrices, *Physical Review Letters* 60, 19, 1895 (1988).

Sornette, D., *Critical Phenomena in Natural Sciences: Chaos, Fractals, Selforganization and Disorder: Concepts and Tools*, Springer Science and Business Media (2006).

Stauffer, D. and Aharony A., *Introduction to Percolation Theory*, CRC press, 1994.

Stein, E. M. and Shakarchi, R., *Complex Analysis* (Vol. 2), Princeton University Press (2010).

Strogatz, S. H., *Nonlinear Dynamics and Chaos: with Applications to Physics, Biology, Chemistry, and Engineering*, CRC Press (2018).

Suzuki, Y. and Dudko, O. K., Single-molecule rupture dynamics on multidimensional landscapes, *Physical Review Letters*, 104, 4, 048101 (2010).

Taleb, N. N., *The Black Swan: The Impact of the Highly Improbable*, Random House (2007).

Theis, C. V., The relation between the lowering of the piezometric surface and the rate and duration of discharge of a well using ground water storage, Washington, DC: US Department of the Interior, Geological Survey, Water Resources Division, Ground Water Branch, 5 (1935).

Touchette, H., The large deviation approach to statistical mechanics, *Physics Reports*, 478, 1–3, 1 (2009).

Tracy, C. A. and Widom, H., Level-spacing distributions and the Airy kernel, *Communications in Mathematical Physics*, 159, 1, 151 (1994).

Tricomi, F. G., *Integral Equations*, Courier Corporation (1985).

Van Kampen, N. G., *Stochastic Processes in Physics and Chemistry*, Elsevier (1992).

Vezzani, A., Barkai, E., and Burioni, R., Single-big-jump principle in physical modeling, *Physical Review E*, 100, 1, 012108 (2019).

Vivo, P., Large deviations of the maximum of independent and identically distributed random variables, *European Journal of Physics*, 36, 5, 055037 (2015).

Walley, P. A. and Jonscher, A. K., Electrical conduction in amorphous germanium, *Thin Solid Films*, 1, 5, 367 (1968).

Wang, W., Vezzani, A., Burioni, R., and Barkai, E., Transport in disordered systems: the single big jump approach, *Physical Review Research*, 1, 3, 033172 (2019).

Wigner, E. P., Characteristic vectors of bordered matrices with infinite dimensions, *Annals of Mathematics*, 62, 3, 548 (1955).

Wigner, E. P., Statistical properties of real symmetric matrices with many dimensions, in: *Proceedings of the Canadian Mathematical Congress*, University of Toronto, pp. 174–184, (1957).

Williams, M., The missing curriculum in physics problem-solving education, *Science and Education*, 27, 3-4, 299 (2018).

Ziff, R. M. and Newman, M. E. J., Convergence of threshold estimates for two-dimensional percolation, *Physical Review E*, 66, 1, 016129 (2002).

Index